Prevention of Reflective Cracking in Pavements

RILEM Report Series

1 Soiling and Cleaning of Building Facades
TC 62-SCF Edited by L.G.W. Verhoef

2 Corrosion of Steel in Concrete
TC 60-CSC Edited by P. Schiessl

3 Fracture Mechanics of Concrete Structures - From Theory to Applications
TC 90-FMA Edited by L. Elfgren

4 Geomembranes - Identification and Performance Testing
TC 89-FMT Edited by A. Rolin & J.M. Rigo

5 Fracture Mechanics Test Methods for Concrete
TC 89-FMT Edited by S.P. Shah and A. Carpinteri

6 Recycling of Demolished Concrete and Masonry
TC 37-DRC Edited by T.C. Hansen

7 Fly Ash in Concrete - Properties and Performances
TC 67-FAB Edited by K. Wesche

8 Creep in Timber Structures
TC 112-TSC Edited by P. Morlier

9 Disaster Planning, Structural Assessment, Demolition and Recycling
TC 121-DRG Edited by C. De Pauw and E.K. Lauritzen

10 Applications of admixtures in Concrete
TC 84-AAC Edited by A.M. Paillere

11 Interfacial Transition Zone in Concrete
TC 108-ICC Edited by J.C. Maso

12 Performance Criteria for Concrete Durability
TC 116-PCD Edited by J. Kropp and H.K. Holsdorf

13 Ice and Construction
TC 118-IC Edited by L. Makkonen

14 Durability Design of Concrete structures
TC 130-CSL Edited by A. Sarja and E. Vesikari

15 Temperature Stresses in Young Concrete
TC 119-TCE Edited by R. Springenschmid

16 Penetration and Permeability of Concrete: Barriers to organic and contaminating liquids
TC 146-TCF Edited by H-W. Reinhardt

17 Bituminous Binders and Mixers
TC 152-PBM Edited by L. Francken

18 Prevention of Reflective Cracking in Pavements:
Edited by A. Vanelstraete and L. Francken

RILEM REPORT 18

Prevention of Reflective Cracking in Pavements

State-of-the-Art Report of RILEM Technical Committee 157 PRC, Systems to Prevent Reflective Cracking in Pavements

RILEM
(The International Union of Testing and Research Laboratories for Materials and Structures)

Edited by

A. Vanelstraete and L. Francken
Belgian Road Research Centre, Brussels, Belgium

Routledge
Taylor & Francis Group
LONDON AND NEW YORK

First published 1997 by E & FN Spon

2 Park Square, Milton Park, Abingdon, Oxfordshire OX14 4RN
52 Vanderbilt Avenue, New York, NY 10017

Routledge is an imprint of the Taylor & Francis Group, an informa business

First issued in paperback 2019

A catalogue record for this book is available from the British Library

Publisher's Note This book has been prepared from camera ready copy provided by the individual contributors in order to make the book available for the Conference.

ISBN 978-0-419-22950-6 (hbk)
ISBN 978-0-367-86591-7 (pbk)

Contents

Contributors - RILEM Technical Committee 157 PRC **xi**

Preface **xiii**

1. Cracking in pavements : nature and origin of cracks **1**
G. Colombier

1.1 Introduction 1
1.2 The various types of pavements or road structures 1

1.2.1 RIGID STRUCTURES 1
1.2.2 SEMIRIGID STRUCTURES 2
1.2.3 FLEXIBLE STRUCTURES 2

1.3 The various forms of cracking in pavements 2

1.3.1 POSSIBLE ORIGINS OF CRACKS 2
1.3.2 SHAPES AND PATTERNS OF CRACKS 3

1.4 Cracking types in various road structures 5

1.4.1 CRACKING TYPES AFFECTING ALL STRUCTURES 5
1.4.2 STRUCTURE-SPECIFIC CRACKING TYPES 6

1.5 Reflection of an existing crack in an overlay 8

1.5.1 LOADS OR STRESSES CAUSING MOVEMENTS OF THE CRACK EDGES 9
1.5.2 NATURE OF CRACK EDGE MOVEMENTS 9
1.5.3 SCHEMATIC DESCRIPTION OF CRACK REFLECTION THROUGH AN OVERLAY 11
1.5.4 DETRIMENTAL EFFECTS OF THE APPEARANCE OF A CRACK AT A PAVEMENT SURFACE 13

1.6 Conclusions 15

1.7 References 15

2. Assessment and evaluation of the reflection crack potential **16**
Molenaar and J. Potter

2.1 Introduction 16

2.2 Nature of cracks 17

2.2.1 TRAFFIC INDUCED CRACKING 17
2.2.2 ENVIRONMENTAL INDUCED CRACKING 19
2.2.3 COMBINED EFFECT OF TRAFFIC AND ENVIRONMENT ON THE CRACKING PROCESS 21

2.3 Factors influencing crack propagation 24

2.4 Guidelines for pavement evaluation 28

2.4.1 PROBLEM IDENTIFICATION 29
2.4.2 QUANTIFICATION OF THE PROBLEM 31

2.5 Summary and conclusions 40

2.6 References 40

3. Crack prevention and use of overlay systems **43**
A. Vanelstraete and A. H. de Bondt

3.1 Introduction 43

3.2 Prevention and treatment of cracks before overlaying 44

3.2.1 LIMITATION OF CRACK FORMATION DURING THE INITIAL CONSTRUCTION PHASE 44
3.2.2 PRE-CRACKING TECHNIQUES FOR NEWLY PLACED CEMENT CONCRETE BASES 45
3.2.3 METHODS USED BEFORE OVERLAYING TO ELIMINATE THE ORIGIN OF EXISTING CRACKS 47
3.2.4 METHODS USED DURING REHABILITATION TO WATERTIGHTOR TO LIMIT THE ACTIVITY OF EXISTING CRACKS 47

3.3 The use of an overlay system 48

3.3.1 DEFINITIONS AND COMPONENTS OF AN OVERLAY SYSTEM 48

3.4 Conclusions 58

3.5 References 59

4. Characterization of overlay systems **61**
A. Vanelstraete, A.H. de Bondt, L. Courard

4.1 Characterization of an interlayer system and its components 61

4.1.1 NONWOVENS, GRIDS, STEEL REINFORCING NETTINGS 61
4.1.2 BITUMEN BASED INTERLAYER PRODUCTS 67

4.2 Characterization of overlay systems 67

4.2.1 ADHERENCE TESTS OF OVERLAY SYSTEMS 67
4.2.2 LABORATORY TESTS OF OVERLAY SYSTEMS UNDER REPEATED THERMAL AND TRAFFIC LOADING 71

4.3 Conclusions 80

4.4 References 81

5. Modelling and structural design of overlay systems **84**
L. Francken, A. Vanelstraete, A.H. de Bondt

5.1 Introduction 84

5.2 Input data for modelling 84

5.2.1 ENVIRONMENTAL AND LOADING CONDITIONS 84
5.2.2 CHARACTERISTICS OF THE BASIC COMPONENTS OF AN OVERLAY SYSTEM 86

5.3 Performance laws 89

5.3.1 THE FATIGUE LAW 89

5.3.2 THE CRACK PROPAGATION LAW 89

5.4 Design Models 90

5.4.1 MULTILAYER LINEAR ELASTIC MODEL 91
5.4.2 APPLICATION OF THE LINEAR ELASTIC MULTILAYER THEORY TO CRACK PROPAGATION PROBLEMS 92
5.4.3 MODELS BASED ON EQUILIBRIUM EQUATIONS 93
5.4.4 MECHANISTIC EMPIRICAL OVERLAY DESIGN METHOD 93
5.4.5 FINITE ELEMENT ANALYSIS 93
5.4.6 THE BLUNT CRACK BAND THEORY 99

5.5 Some remarks about modelling 99

5.5.1 LIMITATIONS OF TWO-DIMENSIONAL MODELLING 100
5.5.2 THE BOUNDARY CONDITIONS 100
5.5.3 UNCONVENTIONAL MECHANICAL PROPERTIES 100

5.6 Conclusions 100

5.7 References 100

6. Implementation of overlay systems on the construction site 104
F. Verhee, J. P. Serfass, T. Levy

6.1 Application of interlayer systems and wearing courses 104

6.1.1 PREPARATORY WORKS 104
6.1.2 SEALING UNDER BITUMINOUS WEARING COURSES OR SURFACE DRESSINGS 106
6.1.3 SAND ASPHALT UNDER BITUMINOUS WEARING COURSES 108

6.1.4 NONWOVENS UNDER BITUMINOUS WEARING COURSES OR SURFACE DRESSINGS 109

6.1.5 MEMBRANE WITH THREADS SPRAYED IN PLACE UNDER BITUMINOUS WEARING COURSES AND SURFACE DRESSINGS 113
6.1.6 GRIDS UNDER BITUMINOUS WEARING COURSES 115
6.1.7 STEEL REINFORCING NETTINGS 116
6.1.8 THREE-DIMENSIONAL HONEYCOMB GRIDS 116
6.1.9 THICK DRESSINGS (SAMIS) UNDER BITUMINOUS WEARING COURSES 119
6.1.10 COMBINED PRODUCTS : GRID ON NONWOVEN 119
6.1.11 OTHER INTERLAYER TECHNIQUES 119

6.2 Wearing courses alone 120

6.2.1 SURFACE DRESSINGS 120
6.2.2 THICK DRESSINGS 120
6.2.3 CONVENTIONAL BITUMINOUS WEARING COURSES 121
6.2.4 SPECIFIC BITUMINOUS WEARING COURSES 121

6.3 Precracking 122

6.3.1 PRINCIPLE 122
6.3.2 EXAMPLE OF PRECRACKING TECHNIQUES 123
6.3.3 IMPLEMENTATION 124

6.4 References 124

7. Summary and conclusions **126**
A. Vanelstraete and L. Francken

7.1 Nature and origin of cracks (chapter 1) 126

7.2 Assessment and evaluation of the crack potential (chapter 2) 127

7.2.1 THE PROBLEM IDENTIFICATION PHASE 127
7.2.2 THE PROBLEM QUANTIFICATION PHASE 128

7.3 Crack prevention and use of overlay systems (chapter 3) 129

7.3.1 PREVENTION METHODS AND TREATMENT FOR TECHNIQUES FOR CRACKS BEFORE OVERLAYING 129
7.3.2 THE USE OF OVERLAY SYSTEMS 130

7.4 Characterization of overlay systems (chapter 4) 131

7.4.1 CHARACTERIZATION OF INTERLAYER PRODUCTS 131

7.4.2 CHARACTERIZATION OF OVERLAY SYSTEMS 132

7.5 Modelling of overlay systems (chapter 5) 132

7.6 Implementation of overlay systems on the construction site (chapter 6) 133

7.7 Remaining issues134

Index **137**

Contributors - RILEM Technical Committee 157 PRC

A.O. Abd El Halim	Center for Geosynthetics, Carleton University, Ottawa, Canada
R. Alvarez Loranca	GEOCISA Laboratory, Madrid Spain
E. Beuving	Center for Research and Contract Standardization in Civil and Traffic Engineering, Ede, The Netherlands
K. Blazejowski	Institut Badawczy Drog i Mostow, Warszawa, Poland
G. Colombier	Laboratoire des Ponts et Chaussées d'Autun, LCPC, Autun, France
L. Courard (Secretary)	Institut du Génie Civil, Université de Liège, Liège, Belgium
A. H. de Bondt	Unihorn bv, Scharwoude, The Netherlands
L. Francken (Chairman)	Belgian Road Research Centre, Brussels, Belgium
H.W. Fritz	EMPA, Dübendorf, Switzerland
W. Grzybowska	Institute of Roads Railways and Bridges, Cracow University of Technology, Cracow, Poland
W.H. Herbst	Central Road Testing Laboratory of the Province of Lower Austria, Spratzern, Austria

T. Levy — Ponts et Chaussées, Division Centrale de la Voirie, Strassen, Grand Duché de Luxembourg

J. Litzka — Institüt fur Strassenbau und Strassenverhaltung, TU Wien, Wien, Austria

R.L. Lytton — Texas Transportation Institute, Texas A & M University, Texas, USA

A.A.A. Molenaar — Faculty of Civil Engineering, Delft University of Technology, Delft, The Netherlands

J.F. Potter — TRL, Berkshire, United Kingdom

J.P. Serfass — SCREG, St-Quentin en Yvelines, France

J. Silfwerbrand — Royal Institute of Technology, Department of Structural Engineering, Stockholm, Sweden

H. Sommer — Forschungsinstitut der Vereinigung der Oesterreichischen Zementindustrie, Wien, Austria

A. Vanelstraete — Belgian Road Research Centre, Brussels, Belgium

F. Verhee — Entr. VIAFRANCE, Rueil Malmaison, France

Preface

The rehabilitation of cracked roads by simple overlaying with a bituminous course is rarely a durable solution. In fact, the cracks rapidly propagate through the new bituminous layer. This phenomenon, called "reflective cracking" is widespread over many countries. It can have many different aspects, in accordance with the large number of factors governing the mechanism of crack initiation and propagation through a road structure.

At a time when the preservation of the road network and adaptation of it to an increasingly aggressive traffic are absolute priorities, road maintenance engineers need to know what is the most appropriate and cost-effective method for treating reflection cracking.

Many solutions have been proposed to meet this challenge :

- preventive measures to be taken during initial construction of the road,
- repair and rehabilitation to be carried out prior to overlaying,
- the use of an appropriate overlay system consisting of an interlayer system and a high performance overlay.

The RILEM Technical Committee 157 PRC "Systems to Prevent Reflective Cracking in Pavements" was formed in 1993 to deal with innovative materials and techniques to prevent reflective cracking. Besides the organization of the third RILEM-Conference on Reflective Cracking in Pavements, the Committee has set up a state-of-the-art-report with the latest experiences about the subject, considering all its different aspects :

- nature and origin of cracks in pavements,
- assessment and evaluation of the crack potential,
- crack prevention and use of overlay systems,
- characterization of overlay systems,
- modelling and structural design of overlay systems,
- implementation of overlay systems on the construction site.

We hope that this work will be a reference for scientists, road engineers and practitioners.

We should like to express our gratitude to all the members of the Technical Committee 157-PRC, especially to A.H. de Bondt, G. Colombier, L. Courard, W. Grzybowska, T. Levy, A.A.A. Molenaar, J. Potter, J.P. Serfass, J. Silfwerbrand, F.Verhee.

Many thanks to all others who contributed to this work, especially to Mrs. A.Thomas of the Belgian Road Research Centre taking care of the lay-out of this report.

A. Vanelstraete
L. Francken
Brussels, June 1997

1

Cracking in pavements : nature and origin of cracks

G. Colombier

1.1 Introduction

Pavements and, more generally speaking, road structures are made up of materials varying very widely both in nature and in properties (unbound aggregates, bitumen-bound materials, materials treated with cementitious binders, ...). All these structures are liable to fracture by many causes, giving rise to cracks of highly different shapes and natures. As a result of traffic and environmental stresses, these different cracks will produce a great variety of stresses in any overlay applied on the cracked structure. Controlling crack reflection through pavements is, therefore, a complex task and any procedure effective for certain types of cracking may be ineffective for others.

A review of all "anti-reflective cracking" systems that have been investigated, experimented with or actually used, requires first of all a clear definition of all the problems to be solved. That is the purpose of this opening chapter.

It is, indeed, vitally important to correctly diagnose the nature and causes of cracks in a structure to be treated, as it is this diagnosis which will direct the choice to proper solutions.

1.2 The various types of pavements or road structures

The most commonly used pavement structures can be classified into three major categories.

1.2.1 RIGID STRUCTURES

These are mainly pavements in internally vibrated or roller-compacted cement concrete. Contraction joints are most often made during construction, by sawing or wet-forming ("preformed joint"). This type of structure is used on all categories of roads, on airfield runways and on industrial and commercial sites.

Prevention of Reflective Cracking in Pavements. Edited by A. Vanelstraete and L. Francken. RILEM Report 18.
Published in 1997 by E & FN Spon, 2–6 Boundary Row, London SE1 8HN. ISBN 0 419 22950 7.

1.2.2 SEMIRIGID STRUCTURES

The base courses of such structures are made of materials treated with smaller amounts of cementitious binder than cement concretes, which results in lower strengths and elastic moduli. As a general rule no joints are made in such courses, which are overlaid with a bituminous wearing course to a thickness varying with the intensity of traffic. These structures are mainly used on medium- or high-volume roads.

1.2.3 FLEXIBLE STRUCTURES

In these structures only untreated materials or materials treated with bituminous binders are used. The thickness of the bituminous courses is generally designed to suit the traffic volumes carried ; it may reach several tenths of centimetres on high-volume roads. On the least trafficked roads the bituminous surfacing is limited to a wearing course, which may be further reduced to a simple surface dressing if the traffic volume is really very low.

1.3 The various forms of cracking in pavements

1.3.1 POSSIBLE ORIGINS OF CRACKS

1.3.1.1 Fatigue
Fatigue due to the pavement having carried a cumulative traffic beyond the limits of its design often results in the appearance of cracks. This "overfatigue" may affect the whole structure (subbase, road base and wearing course) or be limited to the wearing course.

1.3.1.2 Shrinkage
Impeded shrinkage of a pavement layer of virtually infinite length will give rise to cracks as soon as friction with the underlying surface is strong enough to subject that layer to stresses greater than its tensile strength. Shrinkage may be due to the setting of a cementitious binder - when such a binder has been used. It may also be thermal shrinkage, when the temperature in a bound material decreases considerably (day - night, summer - winter). As a rule, shrinkage cracks mainly occur in structures having at least one course treated with a cementitious binder, but in the most severe climates the phenomenon may also affect bituminous courses, especially wearing courses.

1.3.1.3 Movements of the subgrade soil
Movements or a local loss in bearing capacity of the soil on which the structure rests may also cause cracks, which propagate through the various layers of the pavement. This type of cracking may be induced by very different phenomena :

- loss in bearing capacity due to an increase in the moisture content of a poorly drained subgrade,

- slow settlement of a compressible or poorly compacted subgrade under traffic loads and the weight of the pavement,
- land slip, especially along roads with a composite profile (part cut and part fill),
- shrinkage of a clayey soil by excessive loss of moisture during a very dry period. This frequent phenomenon is often exacerbated by the presence of trees along the road,
- frost heave of the subgrade when the thermal barrier formed by the pavement layers is insufficient to prevent frost from reaching a sensitive soil.

1.3.1.4 Constructional defects

Cracking may also originate in certain errors in pavement design or poor practice in constructing one or more courses :

- transverse variation in bearing capacity often occurs at pavement widenings. A longitudinal crack frequently appears at the edge of the old structure, especially when that edge is straddled by the new wheel paths of vehicles,
- construction joints : longitudinal joints between adjacent paving lanes and transverse stop-end joints are weak spots if they have been poorly constructed and no continuity has been ensured in paving. These defects often lead to cracks, both in cementitious and in bituminous materials,
- slippage between layers : when the wearing course is not bonded to the underlying pavement layer, it may rapidly crack under the action of traffic.

1.3.1.5 Ageing and environmental exposure

Cracks starting at the surface of the wearing course can be caused by a combination of thermal contraction and warping of the pavement under cold winter conditions, when the bituminous material is most brittle and least able to accommodate the tensile strain caused by thermal contraction [ref. 1, 2]. This effect increases with time because of age hardening of the binder and the exposure to the environment of the top layer.

1.3.2 SHAPES AND PATTERNS OF CRACKS

Depending on their origin, the type of road structure and the progress of deterioration, cracks may take various shapes and patterns [ref. 3, 4].

1.3.2.1 Orientation

Cracks are most often longitudinal (parallel to the direction of vehicle travel) or transverse (perpendicular to the direction of travel). They are rarely diagonal or parabolic.

1.3.2.2 Shape

The shape of cracks is often relatively rectilinear, but may also be winding or even meandering.

1.3.2.3 Aspect
Cracks may occur as single clean breaks in the pavement ("line" cracks). They may also be double, branched or even interlaced (figure 1.1).

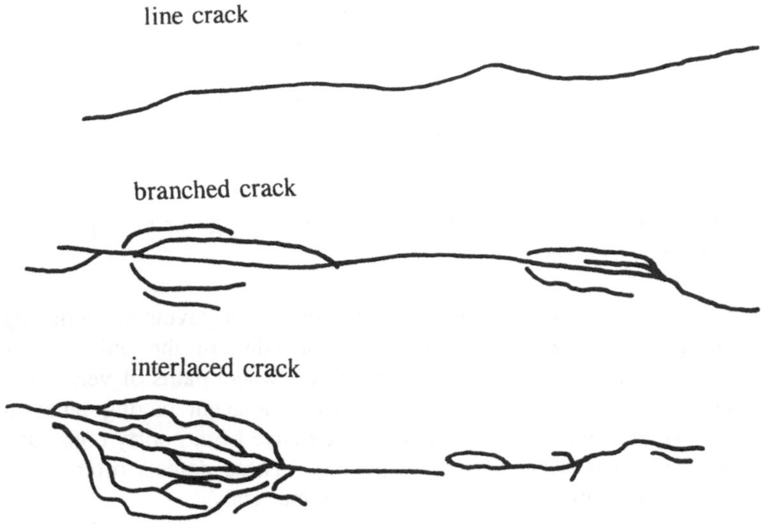

Fig. 1.1 Various aspects of cracks.

1.3.2.4 Width
Depending on the case, the width of cracks (i.e. the distance between the two edges) may be very different. Cracks may be very fine ("hairline" - a few tenths of a mm wide), fine (1 to 2 mm), or wide (several mm up to 1 cm).

1.3.2.5 Pattern
Cracks may be isolated and unconnected or, on the contrary, constitute a more or less dense network or an interconnected pattern of blocks ("block cracking") or small polygons ("alligator cracking" or "crazing") (figure 1.2).

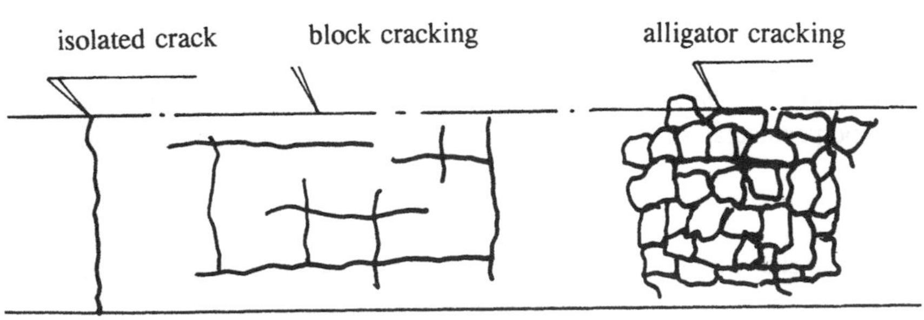

Fig. 1.2 Various crack patterns.

1.4 Cracking types in various road structures

1.4.1 CRACKING TYPES AFFECTING ALL STRUCTURES

1.4.1.1 Cracks caused by the subgrade soil

Loss in bearing capacity and settlement of the subgrade will lead to slab breakage and single longitudinal and transverse line cracks in rigid structures. These cracks will be fine or medium-wide and their edges may stagger in the plane of failure. For semirigid structures the phenomenon may result in cracks of the same type or in block patterns of longitudinal and transverse cracks. In flexible pavements the affected area will craze in the end.

Land slips will cause very wide cracks in any structure. These cracks will follow the slip planes. Their edges will always stagger widely in the plane of failure.

The shrinkage of dried-out clayey soils will induce wide and deep line cracks - either longitudinal or transverse.

1.4.1.2 Surface initiated cracking in bituminous wearing courses

Fatigue cracks in wearing courses are initially fine and limited to the lanes of vehicle travel. With time they extend over the entire pavement surface as generalized alligator cracking.

Cracks starting at the surface of the wearing course can be caused by a combination of thermal contraction and warping of the pavement under cold winter conditions. This phenomenon was largely observed on roads with a lean concrete roadbase [ref. 1, 2]. In cold weather, the upper layers are at a lower temperature than the underlying layers and the consequent differential contraction with depth causes the concrete slab to warp. This effect combined with the thermal contraction in the bituminous surfacing produces a tensile strain in the surface that exceeds the maximum admissible strain of the material and hence a crack appears.

Rapid bitumen ageing at the surface may embrittle the wearing course and make it sensitive to thermal shocks. This results in hairline cracks which propagate from the surface down to the bottom of the layer. This type of cracking may eventually develop into generalized alligator cracking ; the individual cracks will, however, always remain fine.

When poorly constructed (figure 1.3), longitudinal joints between lanes and transverse stop-end joints will open under the action of traffic and variations in temperature. Such constructional defects will lead to line cracks which are often deepened by surface wear and loss of materials.

1.4.1.3 Cracks at widenings

Longitudinal cracks appear where roads have been widened and no continuity has been ensured between the old structure and the widening. Such cracks are generally line cracks and often rather wide.

Fig. 1.3 Transverse construction joint in a bituminous wearing course.

1.4.2 STRUCTURE-SPECIFIC CRACKING TYPES

1.4.2.1 Rigid structures

Construction joints in cement concrete pavements are discontinuities which are similar to cracks. Both longitudinal and transverse joints are very wide at the surface, but they are generally filled with a sealing compound. Nevertheless, they are subject to development in time under traffic and climatic stresses. The edges of the joints may spall off and in particular the joints may loose their capability of transferring loads from one slab to the next ; this may allow relative movements at slab edges ("slab rocking") as vehicles pass over.

When no joints have been made or when they were constructed too late, transverse shrinkage cracks will appear in concrete slabs. These cracks will be relatively rectilinear ; their spacing and width may be very variable.

If the thickness of the concrete slab is inadequate for the traffic to be carried or if the subbase has insufficient bearing capacity, the slab may break under the loads of heavy goods vehicles. The cracks thus formed are medium-wide line cracks (figure 1.4). They may be longitudinal or transverse, or they may affect only corners of slabs.

At the worst the three cracking types described above may appear in one pavement.

Fig. 1.4 Fatigue cracks in a concrete slab (O.C.W. - C.R.R. S5416/5).

1.4.2.2 Semirigid structures

In semirigid structures, base layers treated with cementitious binders normally do not have construction joints. As a result, these layers are subject to natural transverse shrinkage cracking. When reaching the road surface after having propagated through the wearing course, these transverse cracks are most often spaced between 5 and 15 m apart and their width varies with temperature between a few tenths of a mm and several mm (figure 1.5). Shrinkage cracks are often single line cracks when they become visible at the surface, but may develop into double and branched cracks under traffic [ref. 5].

Fig. 1.5 Shrinkage crack in a semirigid pavement.

When base layers treated with cementitious binders are underdesigned or have reached the end of their design service life, they will crack by fatigue. Depending on the residual mechanical characteristics (strength, modulus) of the treated material, cracking in very large blocks will develop into the formation of small slabs or even into generalized alligator cracking.

1.4.2.3 Flexible and bituminous structures

In addition to the cracking types already described for wearing courses, bituminous structures may be subject to cracking by "overfatigue" of the base layers, with the cracks propagating to the wearing course under the action of traffic. Though limited to the wheel paths in their initial stage of appearance, these fatigue cracks generally deteriorate into block cracking (figure 1.6).

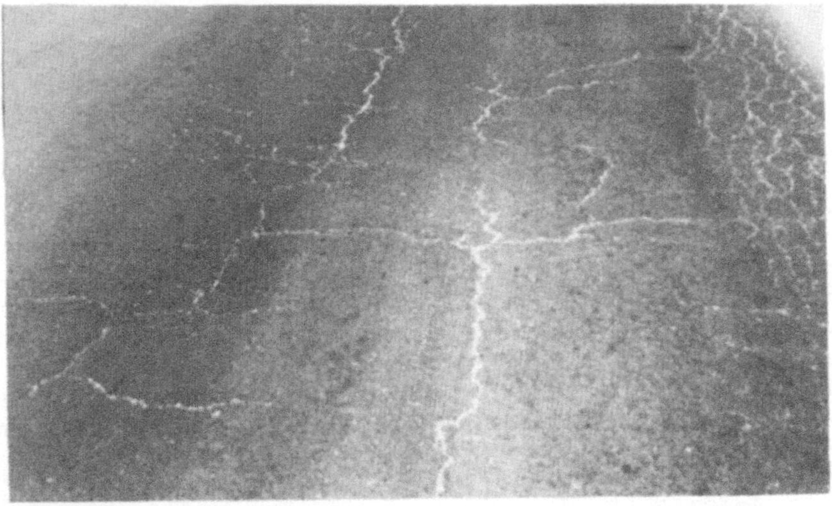

Fig. 1.6 Fatigue cracks in a bituminous pavement.

In countries with very severe winters, bituminous layers may at very low temperatures reach a state of stiffness which makes them liable to crack by thermal shrinkage. The phenomenon is more frequent as hard bitumens or bitumens sensitive to ageing are used. It is then of the same type as described for bases treated with cementitious binders and produces the same result, i.e. regularly spaced transverse shrinkage cracks.

1.5 Reflection of an existing crack in an overlay

The reflection of an existing crack in an overlay is due to the fact that as a result of various factors the edges of an existing crack are subject to movements which are transferred to the bottom of the overlay, where they induce a concentration of stresses. This means that to investigate the problem of reflective cracking it is necessary to define the loads or stresses which may cause crack edges to move, and to analyse the nature of the movements.

1.5.1 LOADS OR STRESSES CAUSING MOVEMENTS OF THE CRACK EDGES

Movements of crack edges may be caused by three sorts of loads or stresses.

1.5.1.1 Traffic

Vehicles, especially heavy goods vehicles, passing over or near a crack press down the loaded edge of that crack. Traffic loading will cause vertical and horizontal movements of crack edges.

1.5.1.2 Variations in temperature

Changes in temperature between day and night or between summer and winter lead to expansions and shrinkages of pavement sections comprised between two successive cracks. These movements will most often result in alternate closure and opening of the cracks. Sometimes a cracked layer may warp under the influence of a severe thermal gradient.

1.5.1.3 Moisture variations in the soil

Moisture changes in the subgrade soil may always induce opening or closure of cracks, whether they initiated the cracks or not.

1.5.2 NATURE OF CRACK EDGE MOVEMENTS

Depending on the structure of the pavement, the nature and shape of the existing crack and the type of loading or stress to which the edges of that crack are exposed, the latter will always exhibit movements with extremely variable characteristics (nature, direction, velocity, amplitude, ...).

1.5.2.1 Type of crack edge movements

The three possible types of movement of the edges of a discontinuity are well known and are classified according to the conventional scheme of G. Irwin [ref. 6] into three modes :

- mode 1 corresponds to opening of the crack,
- mode 2 corresponds to shearing,
- mode 3 corresponds to tearing.

Examples of these types of movement causing the fracture of wearing courses are shown in figure 1.7.

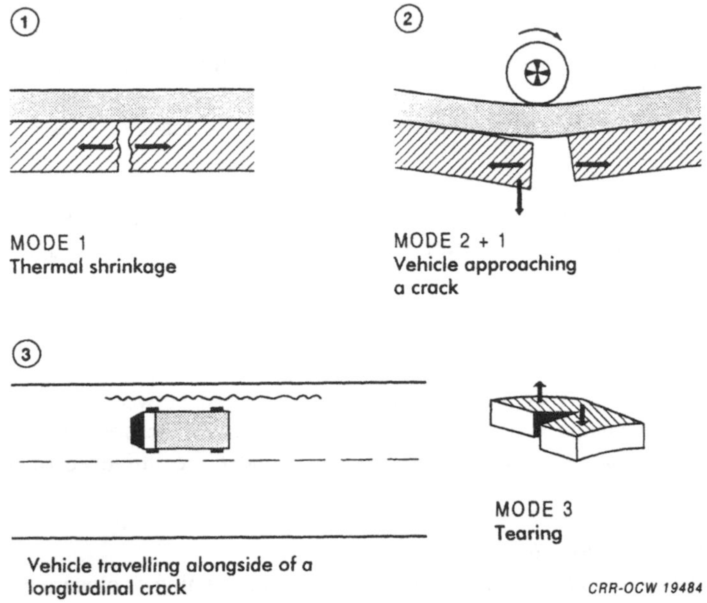

Fig. 1.7 Possible movements of crack edges.

Thermal or drying shrinkage will systematically induce mode 1 crack edge movements (opening). Traffic will cause mode 1, 2 or 3 movements, depending on the position of the vehicle with respect to the crack and also on the geometry of the crack :

- a vehicle approaching a transverse crack will most frequently induce mode 1 and 2 movements. When the axle is straight above the crack, its edges will move in mode 1 (opening),
- a vehicle travelling with its wheel paths astride a longitudinal crack will cause its edges to move in mode 1,
- a vehicle travelling alongside of a continuous longitudinal crack will induce mode 2 movements (shear). At the tip of the same crack (figure 3) the vehicle will cause mode 3 (tearing) movements of the edges.

1.5.2.2 Duration of movement

As far as their velocity is concerned, the crack edge movements resulting from the various possible types of load or stress can be classified into three major categories :

- rapid movements induced by traffic, more particularly heavy goods vehicle traffic. The average duration of movement can be estimated at about one tenth of a second,
- slow movements due to daily variations in temperature (between day and night) and the consequent shrinkage,
- very slow movements resulting from variations between summer and winter (thermal shrinkage) or between a wet and a dry season (drying shrinkage).

1.5.2.3 Amplitude of movements

The amplitude of movement under traffic is clearly a function of the axle loads of the vehicles using the road. It is also directly related to the possible deformations of the structure under traffic. Structural deformation in the vicinity of a crack is a function of the overall bearing capacity of the pavement (which can be assessed for example from deflection measurements), but also depends on whether the structure is capable of total or only partial load transfer across the crack.

For thermally induced shrinkage, the amplitude of movement between crack edges is a direct function not only of the underlying change in temperature but also of the expansion coefficient of the material involved (for example, some limestone aggregates expand only half as much as some siliceous aggregates). The amplitude of crack edge movements due to variations in temperature will furthermore depend on the spacing of cracks and on the quality of the bond between the cracked and the underlying layer.

1.5.2.4 Frequency of movements

The classification for this parameter is similar as for velocity :

- traffic-induced movements are rapid and have a high frequency which approximately corresponds with the number of heavy goods vehicles (a few hundreds to a few thousands per day) using the road,
- movements caused by temperature variations between day and night generally occur twice a day,
- movements related to seasonal variations in temperature or moisture do not occur more than a few times a year.

By way of conclusion :

- traffic-induced stresses in a cracked pavement structure cause crack edges to move in mode 1 (opening), 2 (shear) or 3 (tearing), depending on the position of the vehicle with respect to the crack. The movements are rapid, frequent, and variable in amplitude,
- thermal or drying shrinkage causes crack edges to move in mode 1 (opening). The movements are slow to very slow ; they are infrequent and great in amplitude.

1.5.3 SCHEMATIC DESCRIPTION OF CRACK REFLECTION THROUGH AN OVERLAY

1.5.3.1 The crack reflection process

The development of an existing crack in a superimposed layer under the action of various loads and stresses generally proceeds by three stages involving different mechanisms :

- in the initiation stage a crack is induced by a defect already present in the uncracked layer,

- in the slow propagation stage the crack rises through the full thickness of the layer, starting from the point where it was induced by a concentration of traffic or thermal stresses,
- the fracture or final stage is marked by the crack appearing at the surface of the layer.

The relative importance of the three stages may differ according to the nature of the crack and the type of load or stress acting on the structure. This may be illustrated by discussing the three examples shown in figure 1.8 :

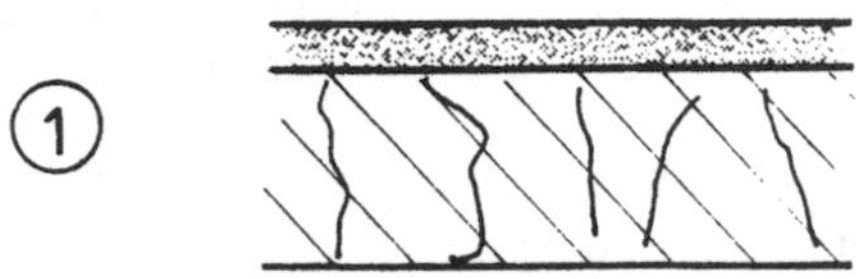

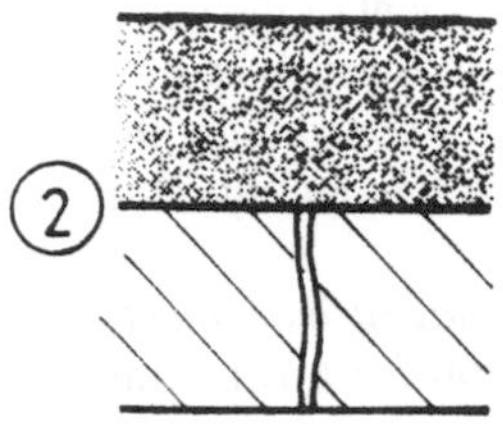

1 : Fatigued structure - Thin overlay - Traffic loading ;
2 : Line crack - Thick overlay - Traffic loading ;
3 : Line crack - Medium-thick overlay - Traffic and thermal shrinkage).

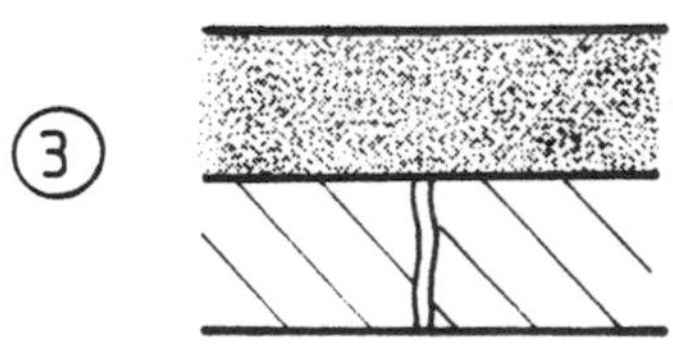

Fig. 1.8 Examples of crack reflection processes.

- example No. 1 : this is a fatigued structure with numerous cracks and a thin (2-cm) overlay. The initiation stage will predominate under the action of traffic and, as soon as the cracks have been induced, they will propagate vertically to the surface,
- example No. 2 : the structure is not subject to fatigue and there are only clearly isolated line cracks. The thick (12-cm) bituminous overlay is perfectly bonded to the cracked layer. The initiation stage is virtually nonexistent, as the crack in the overlay is induced straight above the existing crack because of the strong bond between the two layers. The predominant stage is that of slow reflection of the crack under traffic,
- example No. 3 : the line cracks in a material sensitive to thermal shrinkage have been covered with a 6 to 8 cm thick overlay ; the bond between the two layers is not perfect. The initiation stage will be long, because of the poor quality of adhesion between the layers. Cracks will be induced mainly by thermal shrinkage movements. The subsequent crack reflection stage will be relatively long, owing to the thickness of the overlay. This reflection will be aided mainly by traffic.

1.5.3.2 Crack path
The possible propagation path of an existing crack through an overlay has been described by H. Goacolou and J.-P. Marchand [ref.7] (figure 1.9).

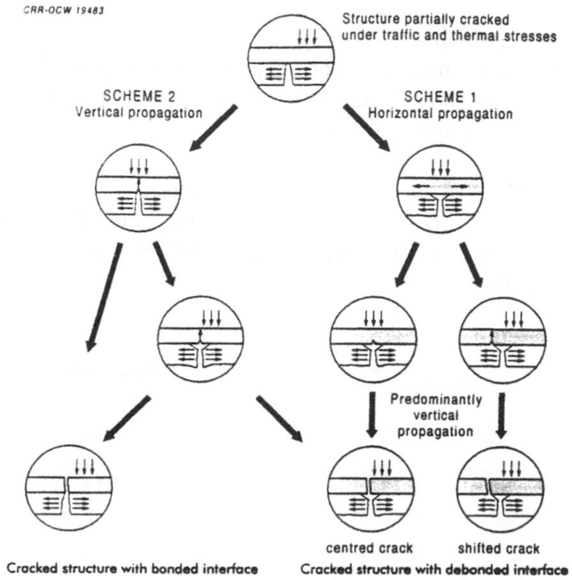

Fig. 1.9 Crack paths. Structure partially cracked under traffic and thermal stresses (Scheme 1 : Horizontal propagation - Predominantly vertical propagation - centred crack - shifted crack - Cracked structure with debonded interface. Scheme 2 - Vertical propagation - Cracked structure with bonded interface).

Under the joint action of traffic and thermal or drying shrinkage, the crack may propagate vertically to the surface without affecting the bond between the layers, or horizontally at the interface while debonding the layers. In the latter case a fatigue process of the overlay is initiated, inducing a new stage of vertical propagation. Such a scheme results in a cracked structure with loss of adhesion between the layers.

1.5.4 DETRIMENTAL EFFECTS OF THE APPEARANCE OF A CRACK AT A PAVEMENT SURFACE

Contract awarders and road managers invariably consider the appearance of a crack at the surface of a pavement as a serious sign of deterioration, with detrimental effects on the behaviour and durability of the pavement.

1.5.4.1 Loss of watertightness
Any crack appearing at the surface is an opportunity for surface water to enter the road foundation and possibly to reach the subgrade soil, which consists of materials that are sensitive to excessive moisture.

1.5.4.2 Stressing of the subgrade

The discontinuity formed by the crack will increase the deformations at the "structural slab" edges under traffic, thus imparting higher stresses to the subgrade soil in that particular point.

1.5.4.3 Increased stresses and strains in the pavement

The aforesaid increase in deformations at the structural slab edges will induce in the road foundation itself - especially in the base layers - severe stresses which will shorten the life of these layers under traffic.

1.5.4.4 Deterioration of the wearing course along the crack

Under the combined action of vehicles, water, frost, etc., the wearing course very often deteriorates by early or late stripping of aggregates or small blocks of asphalt along the crack (figure 1.10).

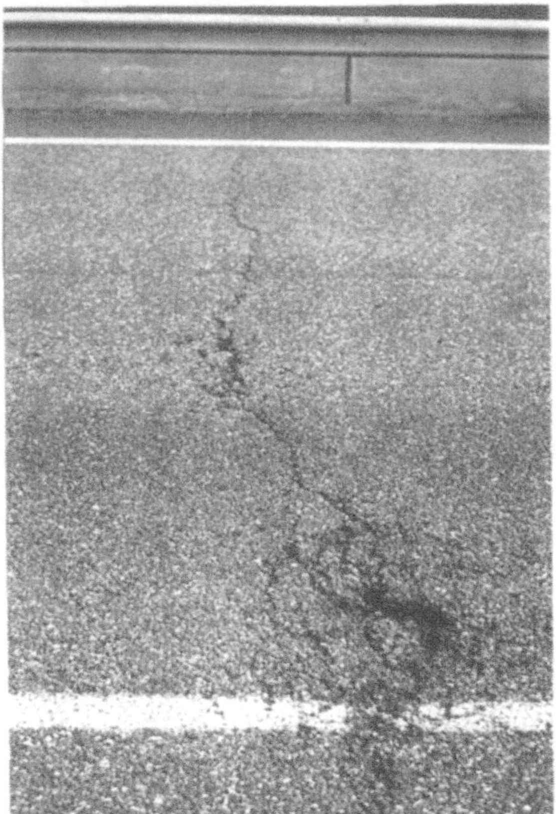

Fig. 1.10 Deterioration of a wearing course along a crack (O.C.W.- C.R.R. S5416/4).

There are methods to repair pavement cracks at the surface, but apart from the fact that these techniques do not eliminate the mechanical effects of the cracks the visual aspect of such repairs does not appeal to road managers and users.
It is, therefore, highly important to have methods and techniques to control the reflection of an existing or potential crack, so that it cannot rise through the wearing course.

1.6 Conclusions

Depending on their nature, the conditions under which they were constructed, and the in-service conditions of loading and stress, the various pavement and road structures may be subject to various forms of cracking. Any cracks appearing at the surface always have a detrimental effect on the pavement. There are ways to at least partly control the reflection of cracks from underlying layers, but the causes of these cracks and the shapes they may take are so varied that there cannot be any universal remedy for them. Before considering any solution to keep reflective cracking under control, one should, therefore, always make a correct and complete diagnosis of the problem to be solved.

It would be desirable to define in which cases the solutions currently known are effective and in which they are not. These matters are discussed in the following chapters.

1.7 References

1. M. Nunn : "An investigation of reflection cracking in composite pavements in the United Kingdom", pp. 146 - 153, 1989.

2. M. D. Foulkes and C. K. Kennedy : "The limitations of reflection cracking in flexible pavements containing cement bound layers. Proc. Int. Conf. on Bearing Capacity of Roads and Airfields, Plymouth.

3. E. J. Yoder, M. W. Witczak : "Principles of pavement design" John Wiley and Sons, Inc. 1975.

4. Catalogue des dégradations de chaussées, SETRA, LCPC, France, 1972.

5. G. Colombier et all. : "Fissuration de retrait des chaussées à assises traitées aux liant hydrauliques". Bulletin de Liaison des LPC, France, n°156 et 157, Juillet et Septembre 1988.

6. G. R. Irwin : "Analysis of stress and strains near the end of a crack traversing a plate". Journal of applied mechanics, Vol. 24.

7. H. Goacolou, J. P. Marchand : "Fissuration des couches de roulement". 5ème conférence internationale sur les chaussées bitumineuses. Delft, 1982.

2

Assessment and evaluation of the reflection crack potential

by A. A. A. Molenaar and J. Potter

2.1 Introduction

Any crack or joint in a pavement tends to reflect through an overlay placed on the cracked pavement (see figure 2.1). The rate at which the reflection process develops depends on the magnitude of the stress concentrations at the tip of the crack or joint, the resistance of the overlay material to crack propagation and the characteristics of the interface between the overlay and the existing pavement. The stress concentrations at the tip of a crack or joint develop as a result of the bending, shearing and tearing actions of traffic loads and tensile and bending actions caused by temperature and moisture movements as well as temperature and moisture gradients.

When an overlay has to be placed, it is important that the nature and causes of the cracks in the existing pavements are known, as well as the type and amount of movements at the crack or joint because they determine to a very large extent the type and size of the stress concentrations.

In the evaluation of cracked pavements much attention should therefore be given to the assessment of the reflection potential of the cracks which are present in the pavement. Reflection potential can be defined as the likelihood that a crack will reflect through an overlay placed on top of it. The reflection potential is controlled by the size of the differential deformations that will occur at the crack after overlaying. This means that the reflection potential of a crack is dependent on the relative thickness of the overlay and the cracked pavement. Thermal changes within the pavement also play a very important part in the occurrence of reflection cracking. Instead of the "reflective potential", the word "crack activity" could be used as a synonym.

In this chapter attention will be paid to the assessment of the activity or reflective potential of cracks and joints in existing pavements.

Prevention of Reflective Cracking in Pavements. Edited by A. Vanelstraete and L. Francken. RILEM Report 18. Published in 1997 by E & FN Spon, 2–6 Boundary Row, London SE1 8HN. ISBN 0 419 22950 7.

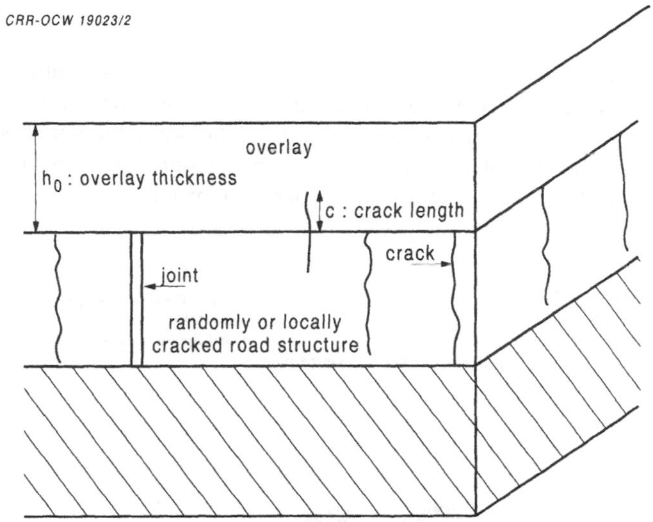

Fig. 2.1 The reflective cracking phenomenon.

2.2 Nature of cracks

An excellent overview of the nature and causes of cracking as well as the reflection of cracks is given in Chapter 1.

The cracks in pavements are predominantly traffic or environmentally induced. A typical example of traffic related cracking is alligator cracking that develops in the wheel tracks. A typical example of environmental related cracking is transverse cracking over the width of the pavement or part of it. This type of cracking is due to contraction when the temperature drops. It is important to recognise the type of cracking and the interaction between the various types of cracks (environmental and traffic). These aspects vary with pavement structures (flexible, semi-rigid or rigid). They are discussed in the later sections of this chapter.

2.2.1 TRAFFIC INDUCED CRACKING

According to classical fatigue theory, cracks due to traffic loads are initiated at the bottom of the bound layers and propagate upwards to the surface. These cracks should occur in the wheel paths and according to theory are transverse in direction. However, a significant amount of longitudinal surface cracking is observed in the wheel paths and it is assumed that they are initiated at the top and have propagated to a depth of approximately 40 to 50 mm. Although the causes of this type of cracking are not well understood it is believed that they could be the result of the non-uniform vertical contact stress distributions at the tyre/pavement interface and the presence of horizontal shear forces lateral to the driving direction.

Figure 2.2 gives an example of the very complex contact stress distributions which were measured under a wide base, or so-called, super single tyre [ref. 1]. This figure shows the vertical contact stresses due to a 50 kN-load and 900 kPa-inflation pressure. The ribs of the tyre can be easily recognised. The maximum contact stress equals 1.31 MPa in this case, which is approximately 1.5 times the inflation pressure. This pattern changes with wheel load, inflation pressure and wear of the tyre.

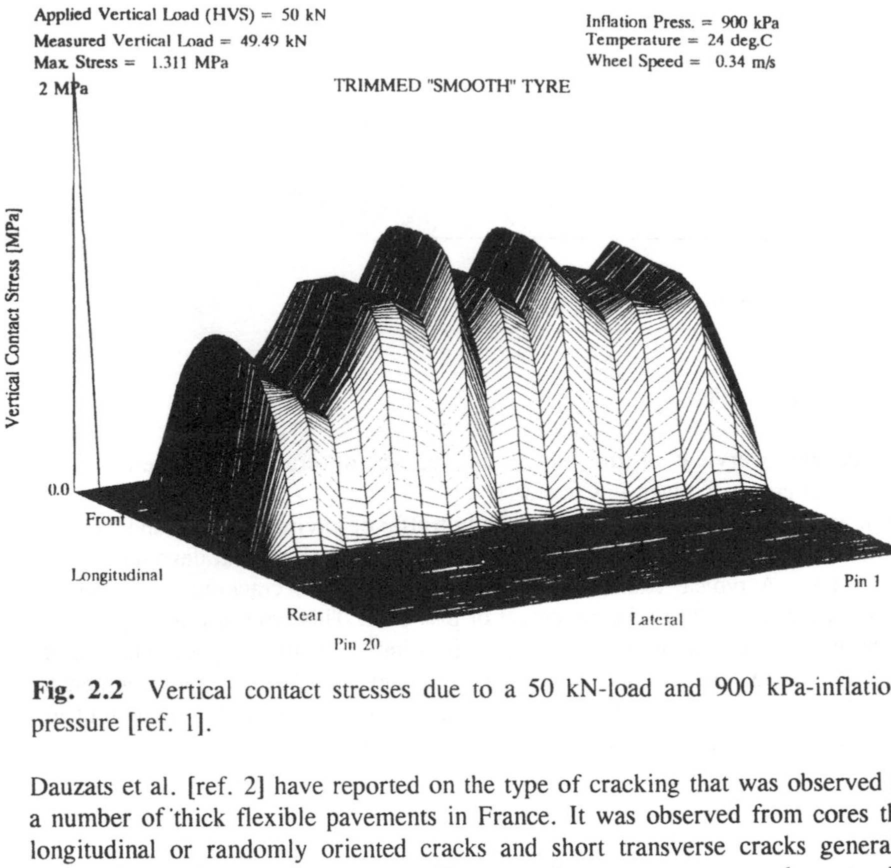

Fig. 2.2 Vertical contact stresses due to a 50 kN-load and 900 kPa-inflation pressure [ref. 1].

Dauzats et al. [ref. 2] have reported on the type of cracking that was observed on a number of thick flexible pavements in France. It was observed from cores that longitudinal or randomly oriented cracks and short transverse cracks generally only affected the wearing course. However, long transverse cracks extended through the whole pavement structure. Of the cores with a crack, 61.5% of the cores taken from a new pavement showed cracking with separation of the top layer. On the remaining 38.5%, the top layer was still fixed to the bottom part of the cores. On the new pavements 39.5% of all the cracks were short transverse cracks, the remaining were longitudinal. Of the cores taken on cracks in overlaid pavements, 55.7% showed cracking with separation and only 9.4% were short transverse cracks.

Of the cores taken on cracks in new pavements only 20% showed cracking from bottom to top. This was also the case for 5.8% of the cores taken on cracks in overlaid pavements. Obviously most of the cracks originated at the pavement surface. Similar observations were made by Nunn [ref. 3].

A similar statement was made by Van Dommelen [ref. 4] who reported that 40% of the maintenance budget for the Dutch motorways is spent on repair of damage due to surface cracks. All this shows that traffic related cracking not necessarily has to develop at the bottom of the bound layers; they can also originate at the top of the pavement.

The question is how to recognise traffic related cracking. In fact this can be done quite easily because this type of cracking will only occur on those places which are loaded by traffic. Of course this type of cracking can propagate to areas which are not subjected to traffic loads but it will stop there because the driving force for crack propagation is very low.

In order to determine the severity of the type of cracking, cores should be taken on cracks to determine the depth of crack propagation and to determine whether or not delamination has occurred between the layers (see 2.4).

2.2.2 ENVIRONMENTALLY INDUCED CRACKING

Environmentally induced cracks are in general transverse in nature because stresses due to shrinkage, resulting from a drop in temperature, are generally highest in the longitudinal direction.

Under special conditions, such as high friction and a large drop in temperature or moisture content, longitudinal cracks can develop and in such cases typical block cracking patterns develop. Generally, environmentally induced cracking is related to the presence of cement treated layers or a heavy clay subgrade with a high plasticity index, which are well known for their susceptibility to variations in temperature and moisture.

Nevertheless, significant temperature stresses can also develop in asphaltic layers, especially in cold temperature regions. In these areas, temperatures can become so low that the asphalt mixtures develop glassy properties which means that brittle fracture is more likely to occur. However cracking can also occur in temperate climates, in spite of the fact that under these conditions stresses can relax quite rapidly in bituminous materials.

Gerritsen [ref. 5] has shown that hardening of the bitumen reduces the stress relaxation capabilities of asphaltic mixtures and for that reason, temperature stresses of significant levels can develop at even moderate freezing temperatures. For example, figure 2.3 [ref. 5] shows the development of thermal stress as a function of temperature and figure 2.4 shows the relaxation of these stresses after having been cooled from 18°C to -5°C. Some characteristics of the mixes are shown in Table 2.1.

Table 2.1 Mix and bitumen characteristics of figures 2.3 and 2.4 [ref. 5].

	Mix number			
	1	2	3	ref
Voids (%)	5.1	2.7	7.2	5.1
Bitumen content (% m/m)	6.3	5.6	6.5	6.1
Pen (25°C)	31	59	36	66
R&B (°C)	57	59.5	56	50

The tests were performed on beams mounted in a frame unaffected by temperature change. The beams were subjected to a cooling rate of 10°C/hour. The samples were assumed to be free of any stress at the temperature at which the beam was placed in the frame (ca. 18°C). One also observes that temperature stress mixes already develop at temperatures of 5°C, this is 13°C lower than the temperature of starting the cooling. Combining the information of figure 2.3 with the information given in Table 2.1, it can be seen that the magnitude of the temperature stress is dependent on the hardness of the bitumen. Mixes with the lowest pen bitumen produce the highest stresses. From figure 2.4 and Table 1 it is clear that the hardening of the bitumen has a large influence on the decrease of temperature stresses with time (relaxation).

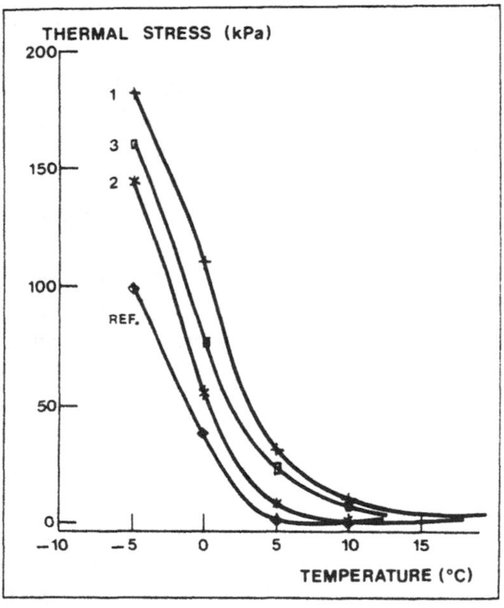

Fig. 2.3 Development of thermal stress in confined asphalt mix samples (cooling rate 10°C/hour) [ref.5].

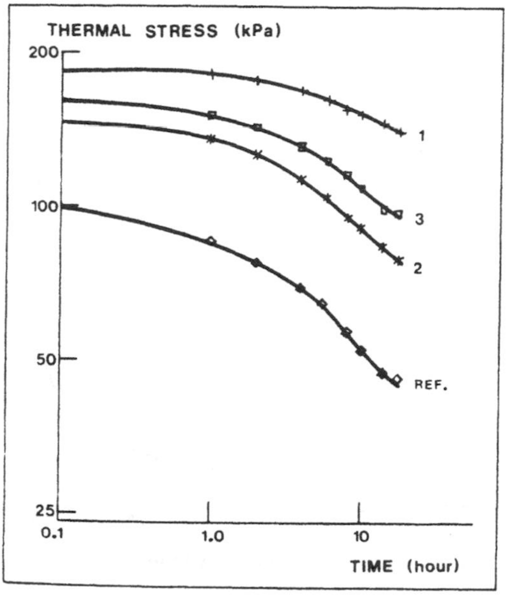

Fig. 2.4 Decrease of the thermal stress with time (relaxation) for a confined asphalt mix sample (temperature -5°C) [ref.5].

Figure 2.4 shows the amount of stress relaxation that occurs when the specimen is kept at -5°C after having been cooled down to that temperature. For the reference mix a relaxation of 50% of the stress is observed (from 100 kPa at the beginning to 50 kPa after 10 hours) while for the hardened mix nr. 1 the relaxation is only 16%.

Although the figures only indicate the effect of hardening of the bitumen, similar trends are observed with mixes containing aggregates with different thermal expansion coefficients.

2.2.3 COMBINED EFFECT OF TRAFFIC AND ENVIRONMENT ON THE CRACKING PROCESS

Traffic and environmentally related stresses do not occur in isolation from each other. Nevertheless, in many climates, traffic induced stresses are dominant during day time while environmentally related stresses (e.g. temperature) are dominant during the night. Goacolou et al [ref.6] and de Bondt [ref.7] have studied the combined effects of traffic and temperature. They showed theoretically that the effect of traffic loads on the development of cracks can be described by the relationships shown in figure 2.5.

This figure shows that the progression of this type of cracking expressed by means of the ratio c/h_0 (crack length ÷ overlay thickness) is a slow process in the beginning, while a very fast progression occurs during the last phase.

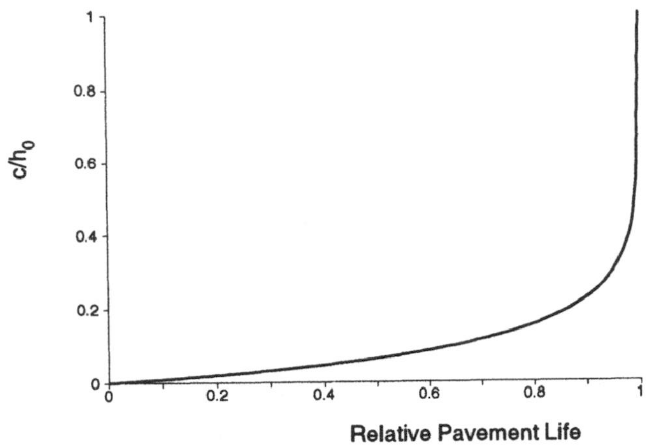

Fig. 2.5 Computational example of reflective cracking caused by traffic loads [ref.7].

The relation shown in figure 2.5 is just an example. The actual shape depends on a number of aspects such as the type of structure and the material characteristics. The example refers to a pavement with an asphaltic top layer on a cement treated base.

Temperature induced cracking can develop in a totally different way. Figure 2.6 shows a computational example of the development of temperature induced cracks for the same pavement used to develop figure 2.5. Temperature induced cracking is fast in the early phase of the cracking process and slows down in the second phase.

By assuming that the traffic induced cracking process dominates during day time and the temperature effects during the night, the overall cracking behaviour is a combination of figures 2.5 and 2.6.

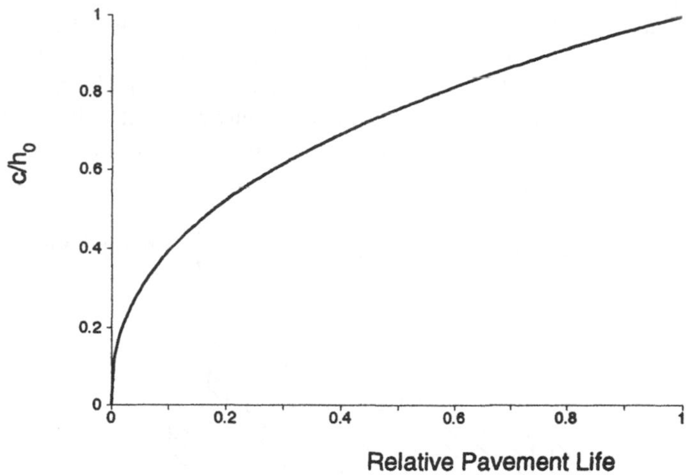

Fig. 2.6 Example of reflective cracking caused by daily temperature cycles [ref.7].

Figure 2.7 is an example of such a combination. In this example the temperature effect was kept constant, but the different amounts of traffic on the various traffic lanes were taken into account [ref.7].

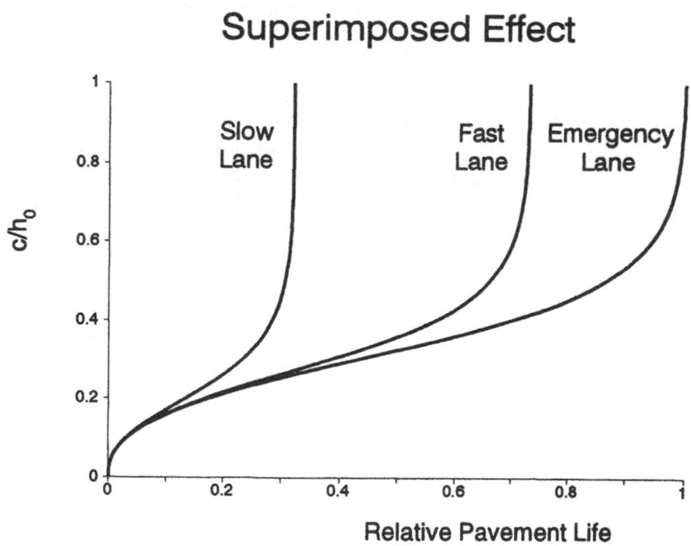

Fig. 2.7 Example of the reflective cracking process for the different lanes of a highway [ref. 7].

This type of behaviour was observed by De Bondt on highway A50. This is illustrated in figure 2.8. The A50 is a main highway in the Northern part of the Netherlands and consists of 200 mm of asphalt on a 400 mm thick sand-cement base layer. Figure 2.8 shows that although a particular type of cracking is usually initiated because of one particular reason (traffic or environment), the combined effects have also to be considered. This should be borne in mind when evaluating cracked pavements and selecting the maintenance strategy.

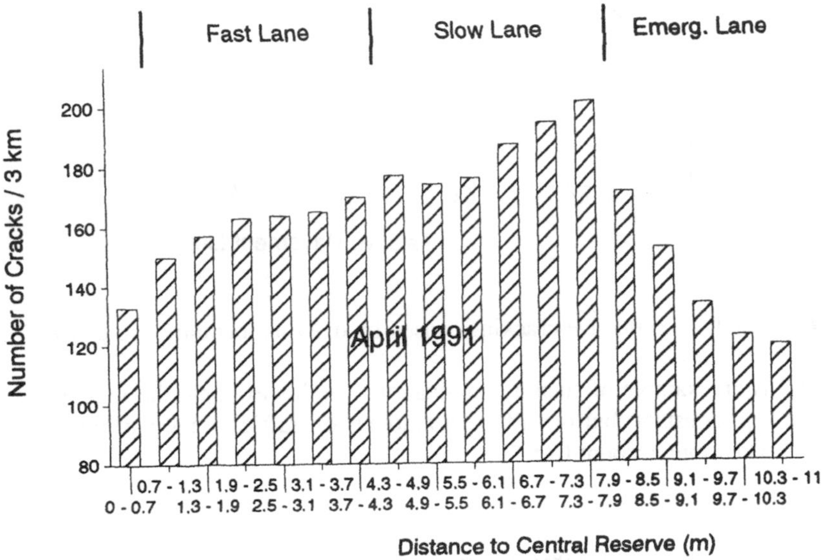

Fig. 2.8 Transverse cracks observed at highway A50 [ref. 8, 9].

In areas with (very) weak subsoils, cracking of the pavement is often initiated by movements in the subsoil, which are settlement or stability related. De Bondt [ref. 7] showed that, in this case, these effects should be analysed in a combined way. He pointed out that cracking due to uneven settlements is accelerated by traffic and vice versa.

2.3 Factors influencing crack propagation

In the previous section, a general description has been given of the nature of cracking and the fact that although cracks may be generated by one specific cause, the combined effect of all loads should be taken into account in order to make a full evaluation and to select the most appropriate maintenance strategy. However, the description was of a qualitative nature and some theoretical aspects should be discussed in order to understand fully the recommendations which will be made regarding pavement evaluation.

Crack growth can be described using Paris' equation [ref. 10] which is :

$$dc / dN = A K^n \qquad (1)$$

where dc/dN = increase of crack length c per load cycle N.
K = stress intensity factor describing the stress conditions at the tip of the crack.
A, n = material constants.

For traffic related cracking, values of n are in the range of 4 to 5.5 ; for temperature induced cracking n lies in the range 2 to 3 [ref. 11, 12]. The parameter A can be calculated using the relation [ref. 12] :

$$\log A = -2.36 - 1.14\, n$$

Typical values for A and n for several mixes are given in [ref. 12].

In order to limit the crack growth, the stress intensity factor K should be low, as well as the material constants A and n. The magnitude of the stress intensity factor not only depends on the magnitude of the load but also on the crack length, the load transfer across the crack and the stiffness characteristics of the existing pavement and the overlay. The influence of the pavement characteristics on the magnitude of K can be defined as the reflective crack potential of the pavement, mentioned earlier.

Finite element analyses using programs like CAPA [ref. 13] are needed to calculate the stress intensity factor. However, for engineering purposes the method developed by Lytton [ref. 14] can be used. This latter method will be discussed because it allows the factors controlling crack reflection to be described simply. Also guidelines with respect to the evaluation of cracked pavements can be generated using this method. Lytton simplified the problem of crack propagation through a pavement to a crack propagating through a beam that is fully supported by an elastic foundation (figure 2.9).

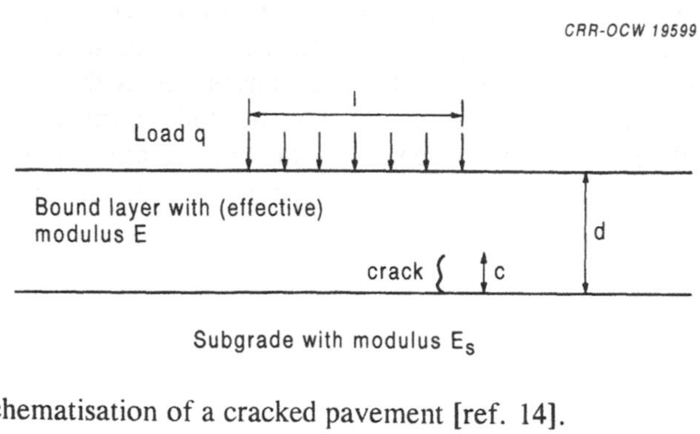

Fig. 2.9 Schematisation of a cracked pavement [ref. 14].

The stress intensity factor at the tip of the crack due to traffic loads (K_{tr}) is dependent on a number of factors (for their meaning : see figure 2.9) which can be written as :

$$K_{tr} = k_{tr} \cdot f(q, l, d, E_s, E) \qquad (2)$$

where k_{tr} is the dimensionless stress intensity factor which is dependent on the ratio c/d, on the amount of load transfer across the crack, and on the type of load that is generated; whether it is bending or shearing.

Figure 2.10 shows the dependency of k_{tr} on the factors mentioned above. This figure clearly shows that the shearing action of traffic is mainly responsible for crack propagation. Furthermore, it shows that, if the overlay is fully bonded to the existing pavement, the ratio of the thickness of the existing bound layers to the thickness of existing bound layers plus thickness of the overlay is an important factor because it is equal to the ratio c/d. Also the influence of load transfer across the crack is clearly indicated.

The stress intensity factor at the tip of the crack due to temperature movements K_T is dependent on the modulus E of the material in which the crack is propagating, its coefficient of thermal expansion α, the change in temperature ΔT, the crack spacing s, the thickness d and the non-dimensional stress intensity factor k_T, as shown in equation (3) below :

$$K_T = k_T \cdot f(E, s, \alpha, \Delta T, d) \qquad (3)$$

Figure 2.11 shows the dependency of k_T on the ratio c/d. Again, it can be concluded that, for fully bonded overlays, the thickness of the existing bound layers in relation to the thickness of the bound layers of the existing pavement and the overlay is an important factor in controlling the rate of crack propagation.

From these theoretical principles, the importance of taking into account the combined effect of traffic and environmental loads can also be seen. For example, in the case of large transverse crack spacings, the thermally induced cracks will be widest at low temperatures. This reduces the load transfer across the crack which results in higher stress intensity factors due to the traffic loads. Furthermore, it can be observed that milling the existing pavements in the cracked areas should have a positive effect on the resistance to crack propagation because the ratio c/d is decreased.

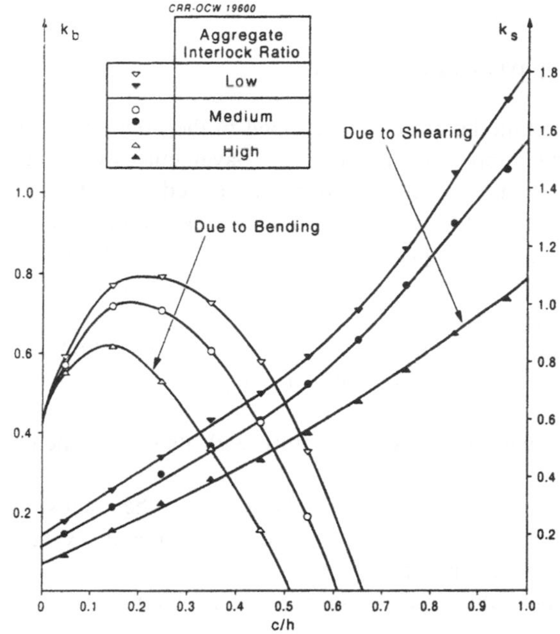

Fig. 2.10 Non-dimensional bending and shearing stress intensity factors versus non-dimensionalized crack length (c/d, where c and d represent the crack length and the combined thickness of the existing pavement surface and overlay, respectively) [ref. 14].

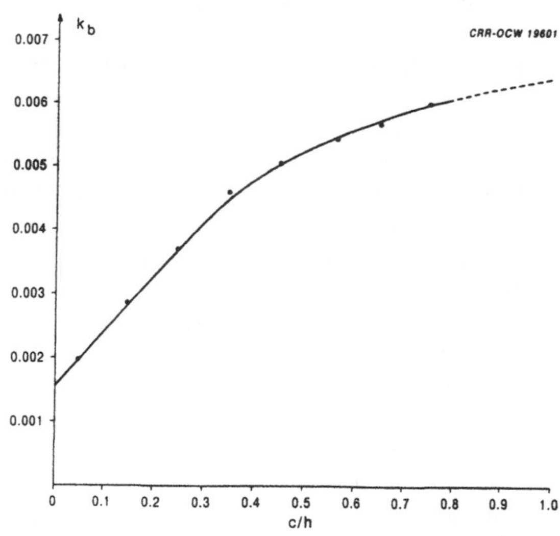

Fig. 2.11 Non-dimensionalized thermal stress intensity factors versus non-dimensionalized crack length (c/d, where c and d represent the crack length and the combined thickness of the existing pavement surface and overlay, respectively) [ref. 14].

2.4 Guidelines for pavement evaluation

Although a simplified model has been described in the previous sections, it can be used to derive some important guidelines for pavement evaluation. It should be noted that the guidelines given hereafter are related to the evaluation of the potential of a crack or joint to reflect through an overlay. The question of whether the crack is occurring in an asphalt pavement with a granular base or an overlaid concrete pavement is in fact not relevant. It is the determination of the potential to crack which is of importance.

The first stage in the evaluation of an existing pavement is the problem identification phase. It involves the following steps :

- Assessment of the type of road structure with its environmental and loading conditions.
- A visual condition survey to identify the nature of the cracks.
- Coring to check the thickness and state of the bound layers and the depth over which the crack has propagated.
- The assessment of the possible causes of the problems.

When this problem identification phase is completed, providing an insight into the possible origins of the problems, a more quantitative and problem-specific evaluation can be performed. This second stage is the quantification phase.

The following measurements are recommended when the main causes are thermal and/or traffic related (other cases, such as frost heave, are not addressed here) :

a. Falling weight deflection measurements (f.w.d) to back calculate the moduli E and E_s. These measurements should be taken on sound areas, and preferably between the wheel tracks.

b. Falling weight deflection measurements to derive the amount of damage due to traffic. These measurements should be taken in the wheeltracks.

c. Deflection measurements at joints or cracks to assess the load transfer across the crack or joint. These measurements should be done together with crack width measurements.

   The measurements at the joints can also be used to estimate whether or not voids or loss of support have developed due to e.g., erosion and pumping. These measurements are especially important in the evaluation of concrete pavements.

d. Crack activity measurements to determine in more detail the degree of slab rocking and the load transfer characteristics across the crack.

e. Crack width measurements together with measurement of crack spacing and changes in crack width because of changes in climatic conditions (temperature, moisture).

It is recommended that measurements are carried out under the most detrimental conditions, i.e., in winter conditions without frost and with high moisture content of the subgrade.

Depending on the nature of the crack, it can be determined whether or not some variation in crack width due to environmental influences is likely to occur. In principle, any environmentally induced crack which is also present in the wheel track area has a high potential for reflective cracking because of the combined effects of traffic and temperature. Coring through such cracks is important in order to establish further the potential for reflective cracking.

If the crack has developed through the entire thickness of bound layers of the existing pavement, then the full measurement assessment program is recommended. If the cracks have not developed through the entire thickness but are limited to say, the wearing course, items c, d and e may be omitted.

With respect to items c and d it is noted that both types of measurements give information on the amount of load transfer across the crack (see 2.4.2).

When the cracks are clearly traffic related, it is not necessary to perform measurements mentioned under e.

With respect to the measurements mentioned under a and b, an f.w.d-deflection survey that is carried out in the wheeltracks gives information on the extent of traffic damage that has developed, which has resulted in loss of bending stiffness. In order to determine the amount of loss in bending stiffness, deflection measurements should also be taken between the wheeltracks because this area should be representative of the original structure. It is not subjected to traffic loads; only ageing should have influenced the pavement condition.

In the next section, the various measurements and their evaluation will be considered.

2.4.1 PROBLEM IDENTIFICATION

2.4.1.1 Assessment of the type of road structure with its environmental and loading conditions

One of the basic requirements for the evaluation of the road structure is a knowledge of the type of road structure, the thicknesses of the different layers, the types of layers and materials that were used, and the history of the road since construction. It is also important to know the condition of drainage and, if possible, that of the base, subbase and subgrade. Some of this information can be provided after coring. Most important is a knowledge of the traffic and environmental conditions to which the road is exposed, with an estimation of the traffic to be expected in the future.

2.4.1.2 Visual Condition Surveys

For overlay design purposes, crack mapping is an important tool because this also helps to determine where the additional measurements can best be taken. This implies that a general assessment in which the condition is rated by means of a single figure (e.g. ranging from 1 good to 5 bad) is not sufficient for this purpose. Also surveys in which the severity and extent of the damage is categorised in classes is not very helpful for the overlay design discussed here. It is not necessary to map the cracks on each section of the pavement to be overlaid ; it can be restricted to areas which are representative of the entire section and to areas where the damage is severe.

Visual condition surveys not only involve mapping of the cracks, but also includes a visual survey of the drainage system. Information is required on the level of the road with respect to that of its surroundings and on the presence or not of large slab stepping or other large unevennesses.

2.4.1.3 Coring

As indicated, cores taken on cracks can give essential information on the potential for reflective cracking and on the condition and thickness of the bound layers. When taking cores, it should be realised that not only the location where the crack is widest at the surface is of interest but also cores taken at the tip of the crack are recommended to provide information on how the crack has developed. This is indicated in figure 2.12. Core A indicates that the crack is through the entire thickness, core B indicates that the crack is probably propagating from top to bottom. Core C indicates that the crack is probably growing from bottom to top. Several cores are sometimes necessary to determine how a crack has developed.

2.4.1.4 Assessment of the possible causes

The evaluation of the road according to the different steps of the problem identification phase, described in 2.4.1.1 to 2.4.1.3, generally makes it possible to assess the reflective cracking potential and to determine possible causes of the cracks. Further measurements are necessary to investigate the existing road and the cracks in more detail and to obtain more quantitative information on certain aspects. This is part of the quantification phase of the problem.

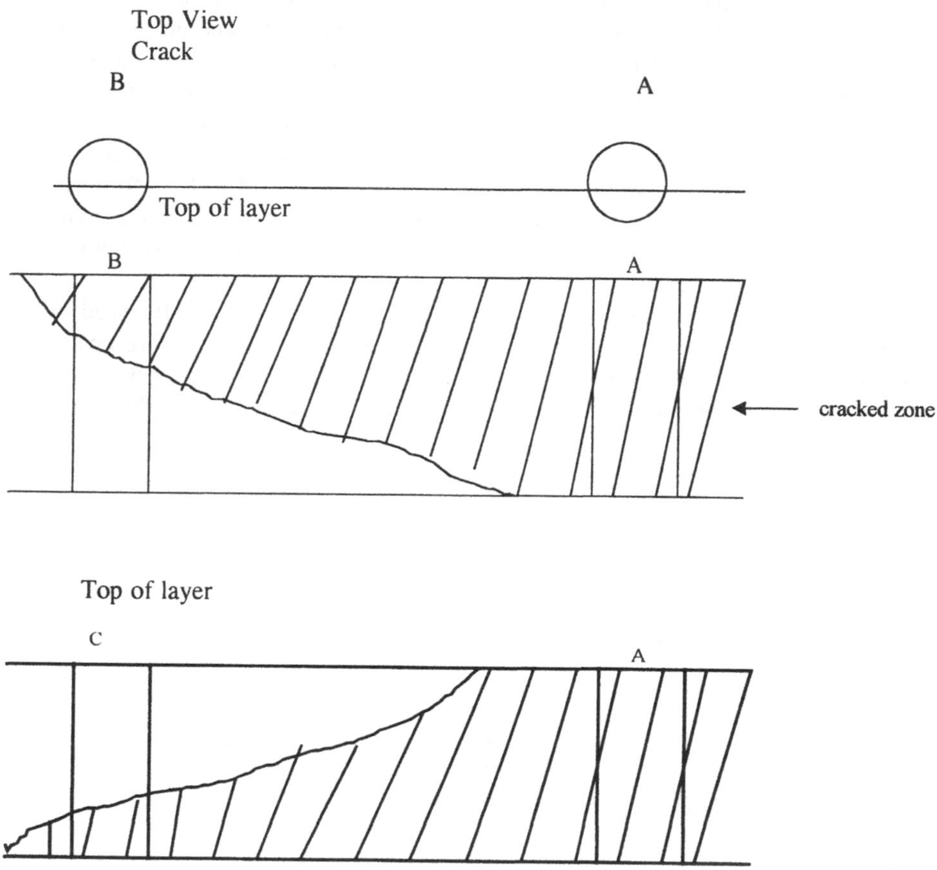

Fig. 2.12 Proposed coring scheme at cracks.

2.4.2 QUANTIFICATION OF THE PROBLEM

2.4.2.1 Deflection measurements to assess damage due to traffic

It is not intended to discuss the measurements of deflection and the analysis to determine the residual life of pavements. The reader is referred to e.g. Irwin [ref. 15], Molenaar [ref. 16], the PIARC Technical Committee on Concrete Pavements [ref.17], Alvarez Loranca [ref. 18], the Belgian Road Research Centre [ref. 19], and Kennedy [ref. 20].

Some of the available methods relate the measured pavement deflection directly to a remaining life using deflection performance charts that are based on the performance of a number of test pavements. Other methods are based on the back calculation of layer moduli from the measured deflection bowls, which is followed by an analysis of the stresses and strains in the pavement using the back calculated layer moduli as input. Emphasis should be placed on the measurement of deflections in and between the wheeltracks because these measurements can be used to compare the condition of the trafficked zone (wheel/track area) with the condition of a zone which is not subjected to traffic loads (area between the tracks), which can be rated as being representative for the original structure.

An example of the differences in deflections that occur in and between the wheeltracks are given in figure 2.13. These data were collected in [ref. 9] during an extensive study of the performance of the A-50 motorway in the Netherlands using the f.w.d. The results were obtained from measurements taken between thermal induced transverse cracks.

The comparison of deflections in and between the tracks is influenced by differences in layer thickness and, for unbound bases, by differences in compaction level. In order to minimise the effect of differences in layer thickness, it is recommended that the deflection measurements are carried out in the pattern shown in figure 2.14.

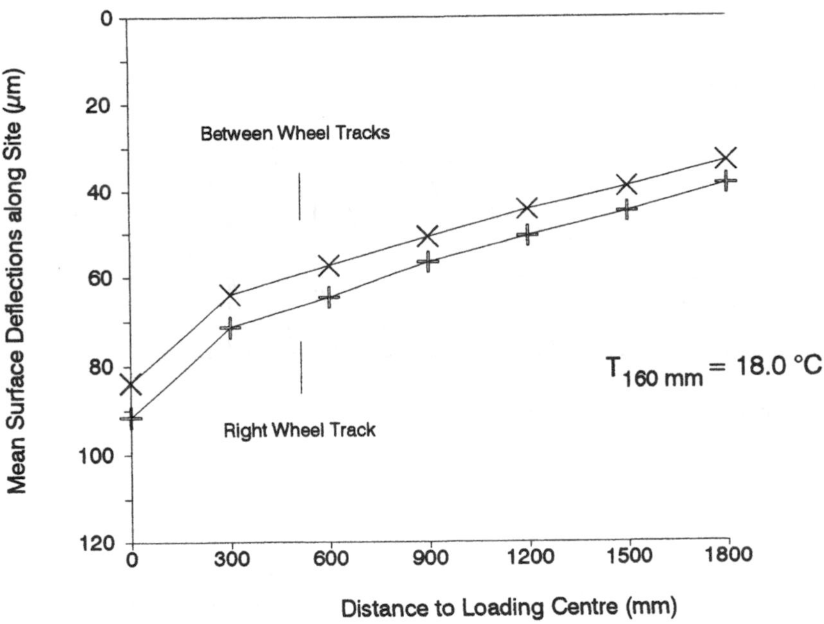

Fig. 2.13 Mean surface deflections at geophone positions between the cracks in the slow lane along the test site for a specific pavement temperature [ref. 9].

The scheme in figure 2.14 can be used in order to allow possible edge effects to be quantified as well.

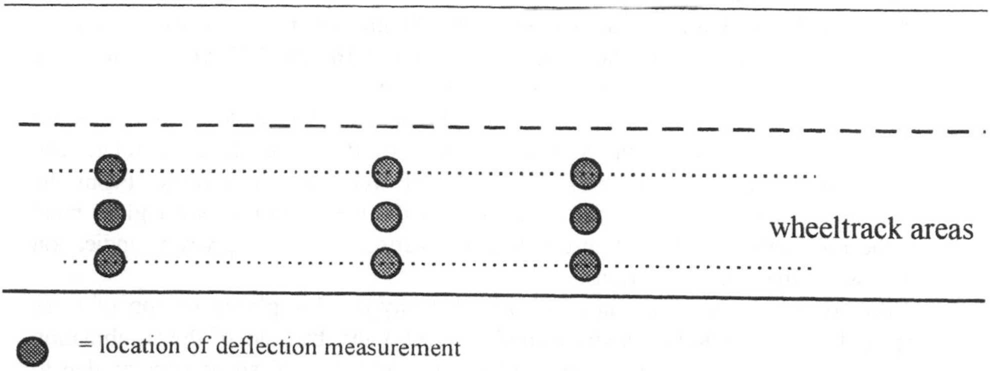

Fig. 2.14 Example of possible locations of deflection measurements between the cracks.

To date, no preference has been given to a particular device for measuring deflection because it is considered that this type of analysis can be done by means of any deflection device provided the measurement protocols belonging to the device are followed.

2.4.2.2 Deflection Measurements to Assess the Load Transfer at Cracks

With respect to the measurement of load transfer across a crack, it is considered that the falling weight deflectometer (f.w.d.) has significant advantages over other deflection devices. For this reason the f.w.d is recommended for load transfer analyses.

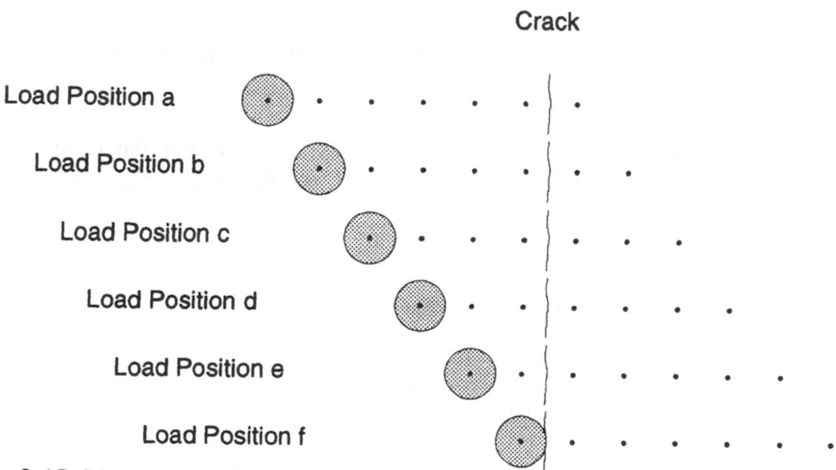

Fig. 2.15 Measurement locations for load position a to f used in [ref. 9].

Figure 2.15 shows an example of the various positions of the falling weight when measuring load transfer. Some results that were obtained in [ref. 8] are shown in figures 2.16 and 2.17. The pavement on which the results were obtained consisted of asphalt layers with a total thickness of 200 mm on top of a 400 mm thick cement treated base on a sand subgrade. Figures 2.16 and 2.17 clearly show the effect of the crack width in winter on the deflections.

When the f.w.d. is in location f, the maximum deflection in the winter is about 1.5 times higher than in the summer because of the wide crack opening. The winter measurements indicate the existence of free edge conditions. From the measurements it is also clear that placement of the f.w.d. in position f gives most of the information, but position e is also interesting because it gives an indication of possible free edge conditions.

Furthermore, figure 2.17 shows clearly that any overlay placed on top of these types of cracks would be heavily loaded during winter, because of the combination of horizontal tensile thermally induced stresses and vertical shear stresses due to traffic. The tensile stresses develop in the overlay because of shrinkage of the cement treated base while the shear stresses are caused by the low degree of load transfer at the crack, due to an increase in the crack width.

It is recognized that the full procedure described in figure 2.15 is not very practicable. Information on the load transfer can also be obtained by using positions as shown in fig. 2.18 or as used in [ref. 21].

The amount of load transfer, LT, is given by :

$$LT = \frac{d_U}{d_L}$$

in which D_U is the deflection of the pavement on the unloaded side of the crack or joint.

D_L is the deflection of the pavement on the loaded side of the crack or joint.

From these f.w.d-measurements two types of approaches can be followed to determine the load transfer efficiency :

- They can be used on an empirical base as in [ref.21, 22]. For the geophone configuration as shown in figure 2.18, research has shown that the following criteria are valuable :

  - good load transfer for $d_U/d_L > 0.9$,
  - satisfactory load transfer for $d_U/d_L > 0.5$,
  - poor load transfer for $d_U/d_L < 0.5$.

- They can be used in finite element analysis.

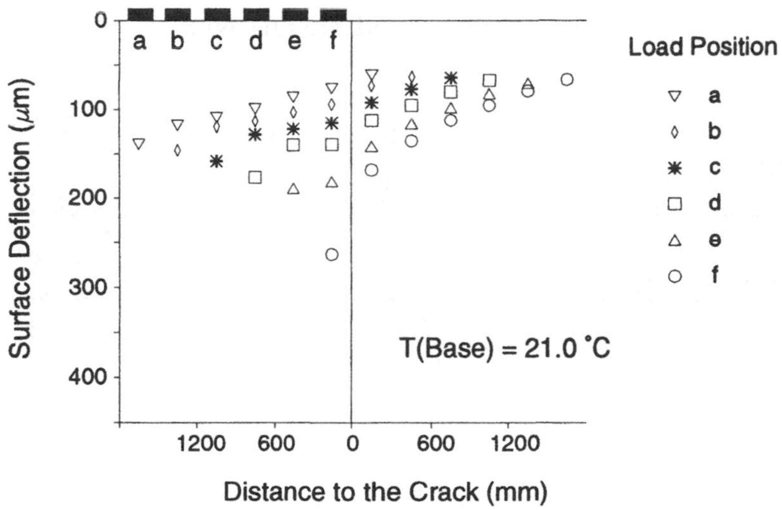

Fig. 2.16 Typical surface deflections in summer at a specific crack [ref. 8].

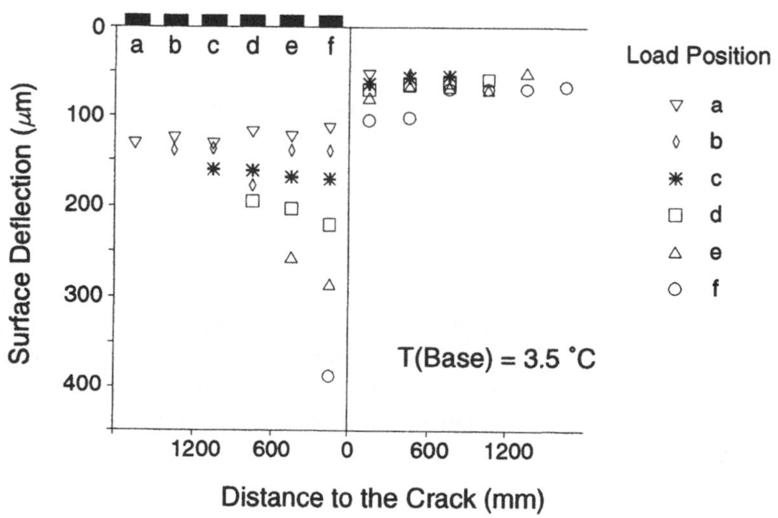

Fig. 2.17 Typical surface deflections in winter at a specific crack [ref. 8].

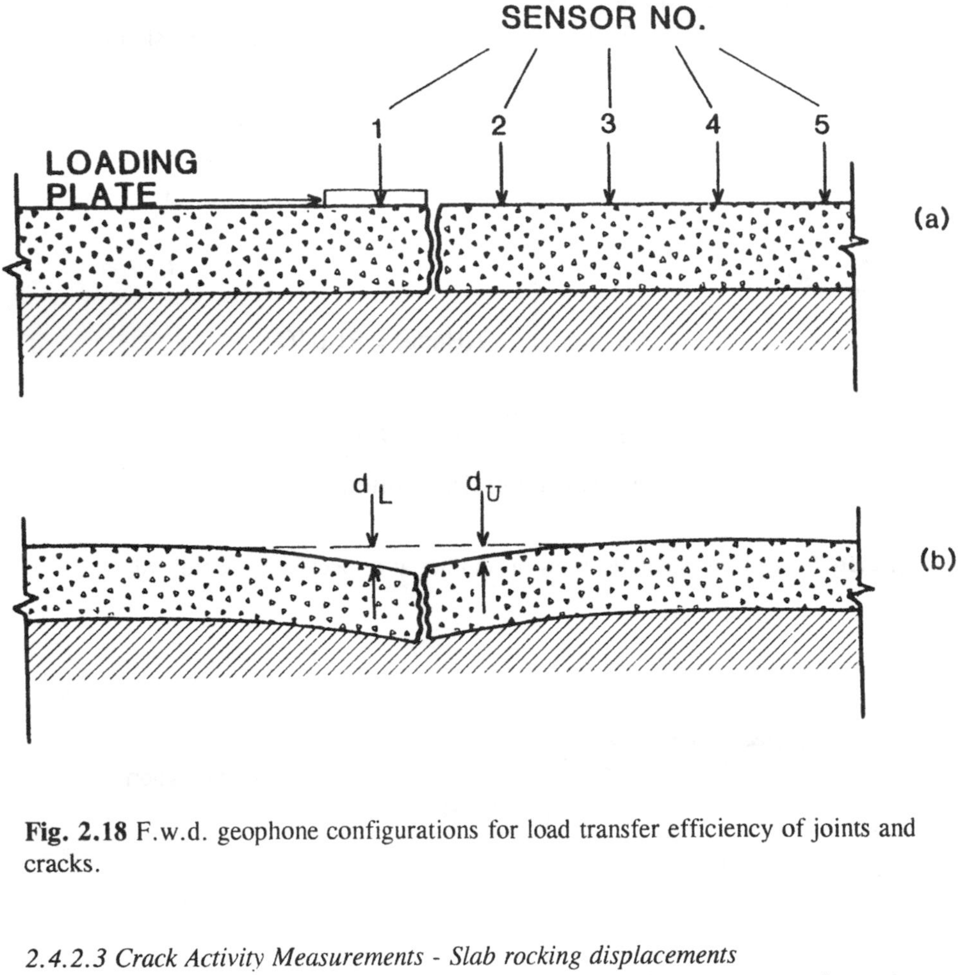

Fig. 2.18 F.w.d. geophone configurations for load transfer efficiency of joints and cracks.

2.4.2.3 Crack Activity Measurements - Slab rocking displacements

As has been indicated in section 2.4.2.2, the load transfer across a crack can be analysed by means of falling weight deflectometer measurements. This procedure is based on the application of an impact load to the pavement. However, traffic applies rolling loads to the pavement, so it is of interest to determine the response of a crack as a result of a passing load. For this, a crack activity meter (C.A.M.) can be used. Several pieces of equipment exist [ref. 8, 23, 24]. Examples are given in figures 2.19 and 2.20.

Fig. 2.19 Example of a device used for slab rocking measurements [ref. 23].

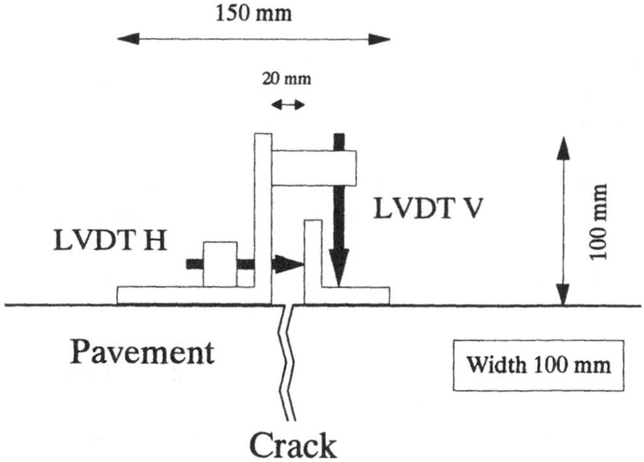

Fig. 2.20 Schematic representation of the crack activity meter used in [ref. 8, 24].

The C.A.M. allows the continuous measurements of both the horizontal and vertical displacement as a result of a passing wheel load. A typical example of the measurement results is given in figure 2.21.

Although measurements of slab rocking are highly dependent on temperature and the moisture content of the subbase and subgrade, they can provide indications as to whether are not the concrete slabs have to be cracked and seated before overlaying. This is discussed in the sixth chapter of this report.

The results of these measurements can also be used in finite element analysis to obtain information about the load transfer across cracks.

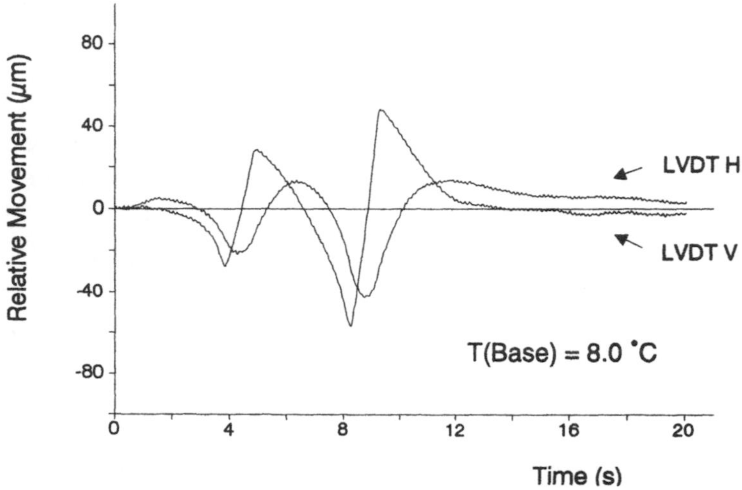

Fig. 2.21 Typical example of load transfer measurement results (truck speed 4 km/h) [ref. 8].

2.4.2.4 Crack Width Measurements in relation to Temperature

As has been indicated earlier, temperature changes affect the width of the crack and the amount of load transfer across the crack when a wheel load is passing over it.

For that reason, the variation of the crack width should be monitored for temperature induced cracks. These measurements can be easily performed by placing a pin on either side of the crack and measuring the variation in distance in relation to the temperature by means of a dial gauge connected to a frame unaffected by temperature. Special attention should be given to the measurements of the temperature. It is recommended that both the air temperature as well as the temperature of the cement treated layer should be measured. In order to do this a small hole should be made in the pavement to the middle of the cement treated layer. It is important that the temperature gradient within the concrete is close to zero. If not, the slab can warp and the measurement of crack width will not be correct.

The importance of these measurements are indicated in figure 2.22 [ref.9], where the change in crack width is shown together with the change in load transfer. This is represented by the ratio d_U/d_L, in which d_L is the deflection under the loading plate and d_U is the deflection at the other side of the crack. The d_U/d_L-values (for load position f of figure 2.15) are indicated by black dots in figure 2.22a. In figure 2.22b the associated deflection curves are shown. This example shows a rapid loss in load transfer with decreasing temperature. This indicates that in this case the maintenance strategy has to be designed in such a way that it is able to absorb, without premature failure, the tensile stresses due to a temperature drop and the shear stresses due to traffic.

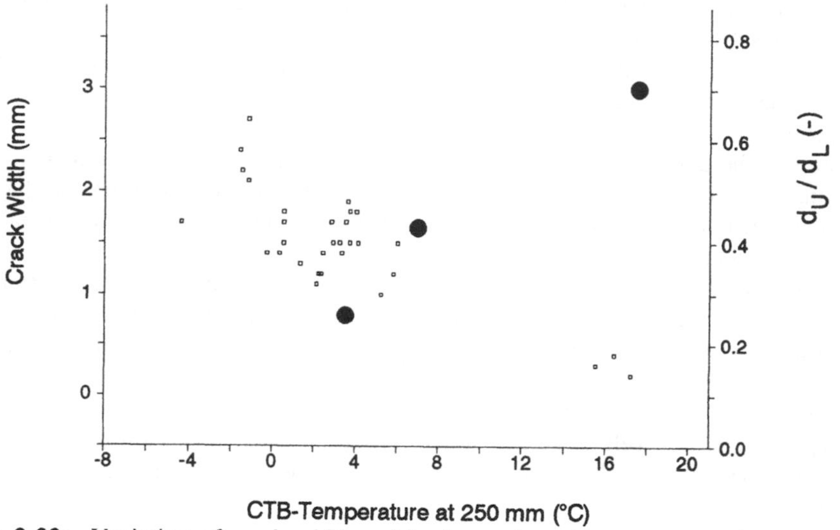

Fig. 2.22a Variation of crack width and load transfer with temperature [ref. 9].

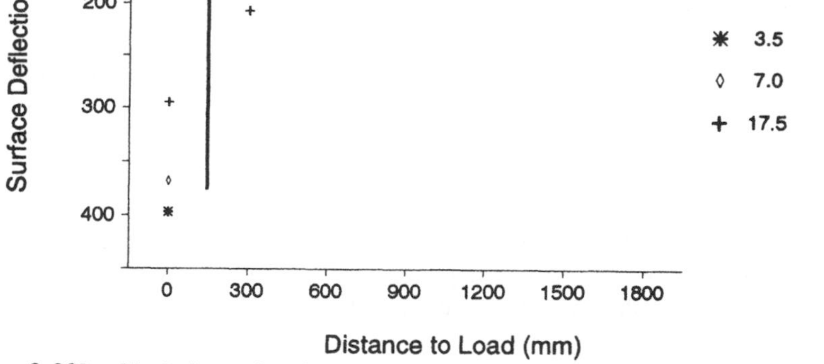

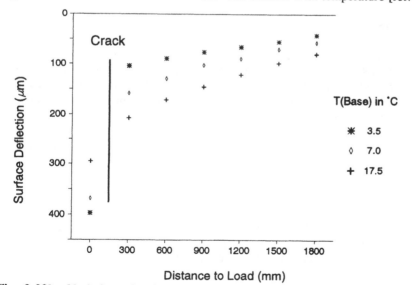

Fig. 2.22b Variation of deflection profile with temperature and distance to the load [ref. 9].

2.5 Summary and conclusions

In this chapter the evaluation of cracked pavements and the assessment of the reflective potential of cracks which are present in the pavement that require maintenance are discussed.

It is shown that cracks can develop for various reasons. A close observation of the crack pattern provides information on the causes and the nature of the cracks. This can be substantiated by means of coring.

Furthermore, it has been shown that the combined effect of traffic and environment on the reflection of cracks through the overlay should be recognised and quantified.

In order to be able to understand which types of measurements are needed for assessing the reflective potential of existing cracks, the crack reflection process has been described by means of a simplified model. The model shows the factors involved and their relative importance.

Based on these considerations, a scheme has been developed for the evaluation and assessment of the reflective potential of various types of cracks in pavements. The measurements involved have been described and suggestions are made for the analysis of the data.

In conclusion, a recognition of the type of cracking and an evaluation of the existing pavement as well as an assessment of the potential reflection of existing cracks is essential in order to be able to design the appropriate maintenance treatment.

2.6 References

1. M. de Beer, J. Groenendijk, C. Fisher : "Three-dimensional contact stresses under the LINTRACK with widebase single tyres, measured with the Vehicle Road Surface Pressure Transducer Array (VRSPTA) System in South Africa", Contract Report CR-961056, Division of Roads and Transport Technology, CSIR, Pretoria, 1996.

2. M. Dauzats, R. Linder : "A method for the evaluation of the structural condition of pavements with thick bituminous road bases". Prod. 5th Int. Conf. Structural Design of Asphalt Pavements, Vol. 1, pp 387 - 409, Delft, 1982.

3. M.E. Nunn : "An investigation of reflection cracking in composite pavements in the United Kingdom". Proceedings of the 1st RILEM-conference on Reflective Cracking in Pavements, Liège, pp. 146-153, 1989.

4. A. Van Dommelen, N.Schmorak : "Analysis of structural behaviour of asphalt concrete pavements in SHRP-NL test sections". Proc. Conference on Road Safety in Europe and Strategic Highway Research Program, Prague, Vol. 4A part 7, pp 125 - 137, 1995.

5. A.H. Gerritsen, C.A.P.M. Van Gurp, J.P.J. Van der Heide, A.A.A. Molenaar, A.C. Pronk : "Prediction and prevention of surface cracking in asphalt pavements". Proc. 6th Int. Conf. Structural Design of Asphalt Pavements, Ann Arbor, Vol 1 pp 378 - 391, 1987.

6. H. Goacolou, J.P. Marchand, A. Mouraditis : "Analysis of cracking in pavements and the computation of the time of reflection" (in French). Bulletin de Liaison de Laboratoires des Ponts et Chaussées, No. 125, 1983.

7. A.H. De Bondt : "Superposition of the individual effect of traffic and environmental loads on the reflective cracking process in asphalt concrete overlays". Report 7-95-203-20; Road and Railroad Research Laboratory, Delft University of Technology, Delft, 1995.

8. A.H. de Bondt and L.E.B. Saathof : "Movements of a cracked semi-rigid pavement structure". Proceedings of the 2nd RILEM-conference on Reflective Cracking in Pavements, Liège, pp. 449-457, 1993.

9. A. H. de Bondt and M. P. Steenvoorden : "Reinforced test sections A50 (A6) Friesland". Report 7-95-209-21. Road and Railroad Research Laboratory, Delft University of Technology, 1995.

10. P.C. Paris and F. Erdogan : "A critical analysis of crack propagation laws". From : Transactions of the ASME, Journal of Basic Engineering, Series D, 85 No. 3, 1963.

11. A.A.A. Molenaar : "Structural performance and design of flexible road constructions and asphalt concrete overlays. PhD-Thesis, RRRL, Faculty of Civil Engineering, Delft University of Technology, 1983.

12. M.M.J. Jacobs : "Crack growth in asphaltic mixes", PhD-Thesis Delft University of Technology, 1995.

13. A. Scarpas, J. Blaauwendraad, A. H. de Bondt and A.A.A. Molenaar : "CAPA : A modern tool for the analysis and design of pavements". Proceedings of the 2nd RILEM-conference on Reflective Cracking in Pavements, Liège, pp. 121-128, 1993.

14. R.L. Lytton : "Use of geotextiles for reinforcement and strain relief in asphalt concrete". Geotextiles and Geomembranes; Vol. 8, No. 3, 1989.

15. L.H. Irwin : "Practical realities and concerns regarding pavement evaluation". Proc. 4th Int. Conf. Bearing Capacity of Roads and Airfields. Minneapolis, Vol. 1, pp 19 - 45, 1994.

16. A.A.A. Molenaar : "State of the art of pavement evaluation". Proc. 4th Int. Conf. Bearing Capacity of Roads and Airfields. Vol 2, Minneapolis, pp 1781 - 1796, 1994.

17. PIARC Technical Committee on Concrete Roads. Evaluation and Maintenance of concrete pavements.PIARC publication; Paris, 1992.

18. R. Alvarez Loranca : "Auscultación de la capacidad portante de firmes semi-rígios" (in Spanish) Rutas, 1994.

19. Belgian Road Research Centre : "Rules of good practice for pavement strengthening with cement concrete" (in Flemish or French). Recommendations - R 62/91; Brussels, 1991.

20. C.K. Kennedy and N.W. Lister : "Prediction of pavement performance and the design of overlays". Laboratory report LR 883 Transport and Road Research Laboratory, Crowthorne, 1978.

21. J. F. Potter and J. Mercer : "Performance of the crack and seat method for inhibiting reflection cracking". Proceedings of the 3rd RILEM-conference on Reflective Cracking in Pavements, Maastricht, pp 483-492, 1996.

22. M.E. Nunn and J.F. Potter : "Assessment of methods to prevent reflection cracking". Proceedings of the 2nd RILEM-conference on Reflective Cracking in Pavements, Liège, pp. 360-369, 1993.

23. A. Vanelstraete and L. Francken : "Laboratory testing and numerical modelling of overlay systems on cement concrete slabs". Proceedings of the 3rd RILEM-conference on Reflective Cracking in Pavements, Maastricht, pp. 211-220, 1996.

24. Description of Crack Activity Meter : "National Institute of Transport and Road Research" (CSIS), Pretoria, South Africa, 1991.

3

Crack prevention and use of overlay systems

A. Vanelstraete and A. H. de Bondt

3.1 Introduction

The appearance of cracks at the road surface is a phenomenon to be largely avoided for the good performance of the road. Cracks in the top layer of a pavement cause indeed numerous problems :

- progressive degradation of the road structure in the vicinity of the cracks due to local overstresses,
- intrusion of water and subsequent reduction of the soil bearing capacity,
- discomfort for the users,
- reduction of the safety.

Two categories of methods are currently used to reduce the appearance of cracks at the road surface :

- The prevention or the treatment of cracks before overlaying.
- The use of an appropriate overlay system.

These two categories of methods may be used in combination in order to achieve the best possible performance.

The first category concerns methods to prevent the formation of cracks during the initial construction phase of the road or to treat existing cracks by choosing road structures, techniques and/or materials which induce less severe or no cracking or make existing or unavoidable cracks less active. An overview of these methods are given in 3.2. They are described in more detail in [ref.1].

The second category of methods concerns the choice of the overlay system itself. It is known that rehabilitation of cracked roads by simple overlaying with a thin bituminous course is rarely a durable solution. Cracks rapidly propagate through the new overlay if no special precautions are taken. To avoid this, special care is necessary in determining the design and the characteristics of the asphaltic overlay itself. In addition, an appropriate interlayer system may be used between the old structure and the new overlay as shown in figure 3.1. The purpose of this interlayer system is to slow down the initiation and propagation of cracks in the overlay by reducing the stresses and strains in the overlay. The interlayer system consists of an interlayer product, e.g., a nonwoven, a grid, a netting ..., with an appropriate fixing system/method to guarantee the adherence with the underlayer. An overview of interlayer systems and a detailed description of the different components of an overlay system is given in 3.3.

Prevention of Reflective Cracking in Pavements. Edited by A. Vanelstraete and L. Francken. RILEM Report 18.
Published in 1997 by E & FN Spon, 2–6 Boundary Row, London SE1 8HN. ISBN 0 419 22950 7.

CRR-OCW 19023/1

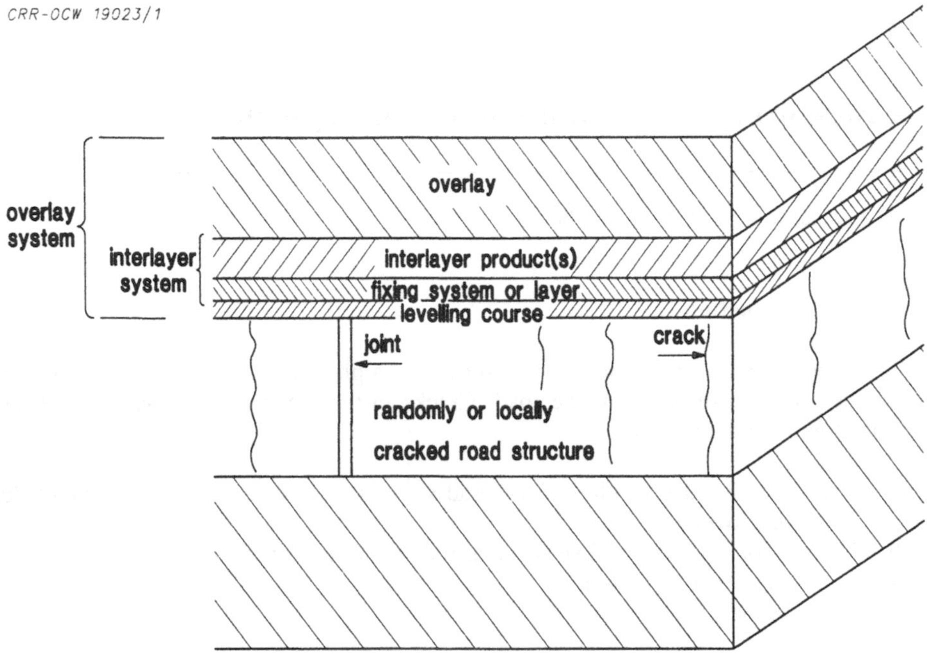

Fig. 3.1 Different components of an overlay system.

3.2 Prevention and treatment of cracks before overlaying

The appearance of cracks at the road surface can already be largely reduced by making use of prevention methods and treatment techniques for cracks before overlaying. A summary is given below. A very good overview for what concerns semi-rigid pavements is also given in the PIARC-report [ref.2].

3.2.1 LIMITATION OF CRACK FORMATION DURING THE INITIAL CONSTRUCTION PHASE

A correct choice of the base materials, a proper design of the road structure, and a good quality of placement are basic rules to respect during the initial construction phase :

- The choice of the materials :

  Binders have to be selected depending on the climatic conditions and the type of mixture. For cement treated bases, it is recommended to use aggregates with low coefficients of thermal expansion, if possible. For bituminous binders, the use of certain polymers or additives can improve their cracking resistance. We shall come back to this in 3.3.1.3.

- The design of the road structure :

  It is clear that roads have to be designed for the traffic levels and temperature conditions to which they are exposed. Insufficient bearing capacity of the road, e.g., by too small layer thicknesses inevitably will lead to an accelerated fatigue cracking. For cement treated bases, uncontrolled cracking should be minimized. We come back to this in 3.2.2.

- The quality of placement :

  The rules of good practice should be followed in laying materials. Defective adhesion between layers and poorly made longitudinal and construction joints cause cracks that can be easily avoided.

3.2.2 PRE-CRACKING TECHNIQUES FOR NEWLY PLACED CEMENT CONCRETE BASES

For roads consisting of one or more layers being treated with hydraulic binders, and thus sensitive to thermal cracking, so-called pre-cracking techniques can be used during the initial construction phase. Their aim is to create more rectilinear and more regularly spaced cracks (generally every 2 - 3 m), which are thus finer and induce smaller crack movements than the natural cracks. Hence, a fast degradation at the crack edges can be avoided and/or the crack propagation through the overlay can be slowed down. Five different pre-cracking techniques are currently used. They are illustrated in figure 3.2.

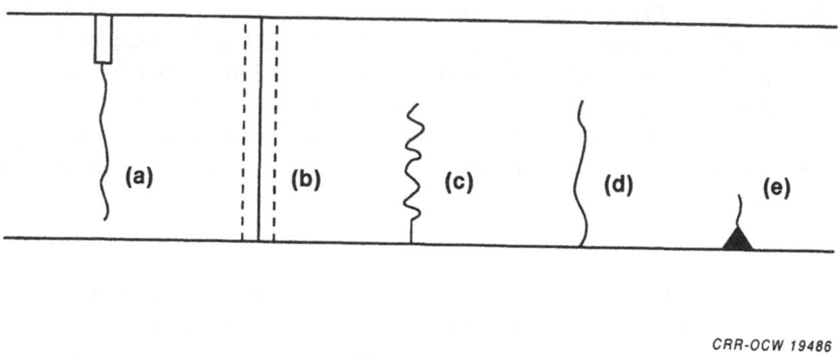

Fig. 3.2 Overview of pre-cracking techniques used in practice.

3.2.2.1 Notches in the upper part of the layer

This method is represented in figure 3.2a and consists of cutting notches in the upper part of the layer after compaction. The notches usually have a depth of about 1/3 to 1/4 of the layer thickness. They can for example be made using a vibrating plate fitted with a knife or by means of a two-wheeled vibrating roller fitted with a cutting disk on one of its wheels.

3.2.2.2 Bitumen emulsion joint

This pre-cracking technique (see figure 3.2b), known as CRAFT (French acronym for Automatic Creation of Transverse Cracks, [ref.3]) consists of cutting a furrow down to the bottom of the layer before compaction, using a blade which makes it possible to spread a rapid-breaking emulsion on the walls of the cut. The furrow is sealed immediately and the layer is then compacted in the normal way. The emulsion creates a low-resistance zone prone to cracking. The bitumen film forms a neat discontinuity and makes the walls of the crack insensitive to water.

3.2.2.3 Insertion of rigid wavy profiles

The technique (see figure 3.2c), known as "active joint" [ref.4] consists of making a groove down to the bottom of the layer after spreading and light compaction. A joint element in plastic material with a wavy profile is inserted, the groove is sealed, and the layer is finished in the normal way. The joint element is 2 m wide or more and placed transversely on the centre line of each lane. Its height is about 2/3rds of the thickness of the layer and it is placed on the bottom of it. The joint forms a crack inducer and by its wavy shape enables loads to be transferred by interlock of the walls of the crack, whatever the underlying material.

3.2.2.4 Insertion of a flexible plastic strip

This technique, also schematically represented in figure 3.2d, and known as "Procédé OLIVIA", [ref.5] consists of introducing in the fresh treated material a flexible plastic strip to introduce cracks. The height of the strip is about 1/3 of the treated material thickness. Its thickness of about 80 μm allows its deformation by the aggregates during compaction ; this results in an efficient load transfer over the whole thickness of the base course.

3.2.2.5 Pre-cracking by initiation of the crack at the bottom of the course

This technique is close to the one described in 3.2.2.1 and consists of initiating the crack by reducing the cross section of hydraulically treated material to induce preferential cracking there. This is achieved by placement of a profile (e.g., triangular wooden laths or cords, see figure 3.2e) at the bottom of the course.

3.2.3 METHODS USED BEFORE OVERLAYING TO ELIMINATE THE ORIGIN OF EXISTING CRACKS

If the origin of the cracks is known, techniques exist in some cases to eliminate the origin of existing cracks before overlaying. If these solutions are technically and economically feasible, they have to be preferred above others, as they are most efficient. Some examples are :

- The cracks are due to a loss in bearing capacity of the subgrade soil by excessive moisture. In this case it is possible to dewater the soil by drainage and to prevent further water ingress by sealing the surface ;
- The cracks are due to general fatigue of the structure. Well-designed structural strengthening will dispose of this problem ;
- The fatigue cracks in the wearing course are due to slippage on the layer beneath. In this case the wearing course can be removed by planing, and a new course can be laid with good bond to the underlying layer.

3.2.4 METHODS USED DURING REHABILITATION TO WATERTIGHT OR TO LIMIT THE ACTIVITY OF EXISTING CRACKS

If the existing cracks cannot be eliminated because of technical and/or economical reasons, preliminary works before overlaying may be necessary in order to limit the activity of existing cracks, even in the case of applying an interlayer system. Some examples are :

- Crack and seat techniques on cement concrete slabs in the case of large vertical movements measured at the joint or crack edges. These techniques limit the activity of cracks.
- Sawing of cement concrete slabs.
- Sawing and sealing of the overlay above the joints in the cement treated bases.
- Injection with cement mortar or epoxy resins to limit the movements of concrete slabs and to fill existing holes under the cement concrete slabs.
- Injection of cracks or joints with bitumen or modified bitumen. This technique prevents water to penetrate to the underlying structure.
- Repair of cracks by bridging them over a width of 10 to 20 cm with a 1 to 2 mm thick elastomer-bitumen film (see figure 3.3). This method is mainly efficient to restore watertightness, but has no effects on the cracking process. It is clear that for the latter two cases, proper installation is of primary importance to guarantee the adhesion between the bitumen film and the existing pavement.

  However, it is important when bridging cracks that good surface texture of the repair is maintained for safety purposes.

Fig. 3.3 Cracks repaired by bridging.

3.3 The use of an overlay system

3.3.1 DEFINITIONS AND COMPONENTS OF AN OVERLAY SYSTEM

The term "overlay system" (see figure 3.1) is used to describe the combined system of a bituminous overlay, interlayer system and levelling course, placed on an underlying road structure. One or more of the components may be absent depending on the quality of the old road structure, the loading conditions, and on the type of rehabilitation system which is chosen. The overlay system is placed on the cracked road structure, usually after prior repairs. They were partly described in 3.2. Other examples are given in chapter 6. The success of using an overlay system for the prevention of cracks depends on the performance of its individual components and their combined performance as a system, being part of a given road structure.

3.3.1.1 The levelling course

A levelling course is a bituminous layer with a mean thickness of a few cm, used on an uneven old road surface in order to obtain a flat surface before placement of the interlayer system. The levelling course generally consists of a dense bituminous mix with maximum aggregate size of about 7 mm.

3.3.1.2 The interlayer system

An interlayer system consists of an interlayer product, fixed on the underlayer with a specific fixing layer and/or placement method depending on the type of interlayer product.

a. Interlayer products

Although the range of commercially available interlayer products is very wide, the large variety of products can be classified in a limited number of categories[1]. Often used are sand asphalt, SAMIs, nonwovens, grids and steel reinforcement nettings :

- Sand asphalt :

  This comprises a thin (10 to 20 mm thick) flexible layer in bituminous sand rich in binder. Sand asphalt used for that purpose is made with sands with grading varying between 0/2 mm and 0/6 mm. Although pure (pen. 80/100 or 180/200) bitumens are used in some cases, sand asphalts are most often applied with polymer modified binders. The range of modifiers is very wide (SBS or EVA copolymers, reticulate styrene-butadiene, powdered rubber). Compositions with 10 to 15 % of fines (< 0.08 mm) and 8 to 12 % of binder are most commonly used ;

- Stress absorbing membrane interlayers (SAMIs) :

  SAMIs (see figure 3.4) consist of a layer of binder (generally modified) which is applied at a high rate (2.5 kg/m^2 on average) and most often spread with single-sized chipping which are rolled in ;

- Nonwovens :

  They are nonwoven plastic (polypropylene or polyester) fabrics not thicker than a few millimetres, which are applied on an existing pavement and saturated with pure or modified bitumen (see figure 3.5). Some of them are manufactured on site [ref.6] ;

- Steel reinforcing nettings :

  These are galvanized steel wire assemblies, which are reinforced at regular intervals with transverse twisted steel wire strands (see figure 3.8).

- Three-dimensional steel honeycomb grids :

  They consist of hexagonal basic elements with a 30 mm thickness. They are tied together with steel rods, perpendicular to the pavement longitudinal axle, nailed to the existing pavement and filled with asphalt. (see figure 3.9 and [ref.8]) ;

The definitions given further for the several categories of interlayer products are slightly different from the ISO- and CEN- definitions of similar products used for geotechnical applications : nonwoven geotextile, woven geotextile, geogrid and geonet. The role they have to play in an overlay (see 3.1.2.4) differs also largely from that in soil and drainage applications.

Fig. 3.4 Chipped SAMI (OCW - CRR 3329/4A).

Fig. 3.5 Nonwoven on sprayed binder (OCW - CRR S4511).

- Grids [2] :

  They consist of regular grids of entirely or almost entirely connected ribs (figure 3.7). These products have visible openings and are made of polypropylene, polyethylene, polyester or glass fibres ;

- Combined products :

  Some interlayer products consist of a combination of two types of products, e.g., grid fixed on nonwoven (see figure 3.10).

---

2 Some authors make a distinction between grids, fillets (nets) and wovens [ref.7] :

- Grids are entirely connected elements, of which the ribs make a fixed angle with each other and with firmly connected intersection points (see fig. 3.6a) ;
- Fillets consist of elements with firmly connected ribs, but which do not make a fixed angle with each other (see fig. 3.6b) ;
- Wovens consist of elements which are not entirely connected. The intersections of the ribs/strands are not really linked (see fig. 3.6c). We note that they have to be distinguished from woven textiles, because of their visible openings. Woven textiles are not considered as products suitable for prevention of reflective cracking here, because of their insufficiently large mesh size on one hand, and too small quantity of absorbable binder on the other hand.

This distinction is mainly made because of the differences in force distribution between these systems. As a result of the entirely fixed ribs in grids, a force exerted on one rib introduces tensile forces on the other ribs. This is also the case for fillets, however to a less extent. In wovens, there is no real force distribution to the other ribs.

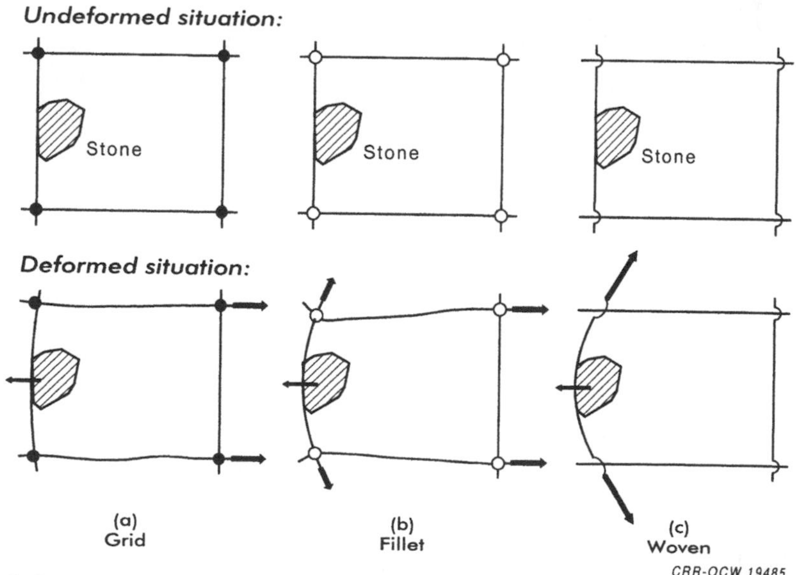

Fig. 3.6 Schematic representation and force distribution for : (a) : grid ; (b) : fillet ; (c) : woven.

Fig. 3.7 Grid in polypropylene (OCW - CRR S 4510).

Fig. 3.8 Steel reinforcing netting (OCW - CRR 3290/10A).

Fig. 3.9 Three-dimensional steel honeycomb grid [ref.8].

Fig. 3.10 Combined product : grid fixed on nonwoven (O.C.W. - C.R.R. S 5419/3).

b. Fixing layers/methods
Different fixing layers or methods, e.g., tack coat, binder layer, slurry seal or nailing are used to provide a good adhesion of the interlayer product to the underlayer. The choice of a given fixing method/layer depends on the type of interlayer product :

- Tack coat :

  This layer is used to fix sand asphalt and grids to the under- and overlayer. It consists of an emulsion, preferably with elastomers. The spreading amount varies from 300 to 500 g/m², according to the surface texture of the supporting layer. It is spread out on the underlayer by preference before placement of the interlayer product.

- Binder coat :

  This layer of bitumen, preferably with elastomers, is used with nonwovens. Its role is threefold : to fix the nonwoven to the under- and overlayer, to impregnate the nonwoven with binder, necessary for its good functioning as anti-cracking system, and for waterproofing. Binder quantities vary from 700 to 1400 g/m² depending on the non-woven product. In some cases, they are applied under the form of emulsion.

- Slurry seal :

  This layer is prepared with sand, cement and emulsion (preferably with elastomeric binder). It is used in combination with a particular steel reinforcing netting and spread out after placement of the interlayer product.

- Nailing :

  Nailing can be used as fixing technique for grids and steel reinforcing nettings, but it is time-consuming and its efficiency largely varies with the nature of the supporting layer. Therefore, nailing is more often used at particular places and in combination with other fixing methods/layers, such as, at the beginning and end of the roll or at places of overlapping (e.g. in curves).

- Self-adhesive :

  Some interlayer products are self-adhesive and do not need any fixing system. This is very practical, however, experience in the field has shown that some of these products loose part of their self-adhesivity when applied on a wet road surface.

c. Combination of interlayer products and fixing layers/methods
Table 3.1 gives an overview of existing combinations of different types of interlayer products with their corresponding fixing methods/layers. Detailed descriptions of their laying procedures are given in chapter 6.

Table 3.1 Existing combinations of different types of interlayer products with their fixing methods/layers.

Fixing method/layer	Tack coat	Binder layer	Slurry seal	Nailing	Self-adhesive
Sand asphalt	X				
SAMI					X
Nonwoven		X			
Grid	X			X	X
Steel reinforcing netting			X	X	
Three-dimensional steel honeycomb grid				X	
Grid on nonwoven	X	X			

d. Role of interlayer systems in the road structure

The role of an interlayer system in the road structure depends in the first place on the type of interlayer system. Its role can be :

- To take up the large localised stresses in the vicinity of cracks and, hence, reduce the stresses in the bituminous overlay above the crack tip. The product in that case acts as reinforcement product. This is the case for grids and steel reinforcing nettings.
- To provide a flexible layer able to deform horizontally without breaking in order to allow the large movements taking place in the vicinity of cracks. This is the case for impregnated nonwovens, for SAMI's and for sand asphalt. This function is also often described as "controlled debonding". It is obvious that total debonding has to be avoided in all cases, otherwise fatigue cracking may appear already very shortly after rehabilitation.
- To provide a waterproofing function and keep the road structure waterproof even after reappearance of the crack at the road surface. This is often the case for nonwovens and SAMI's.

A given product will act as a reinforcement product if its overall stiffness modulus is higher than that of the bituminous overlay. This depends on the type of interlayer product and, on the temperature in the actual road structure. The bituminous overlay material is highly temperature susceptible. Therefore, a given product can be reinforcing in medium and high temperature ranges, but not under winter conditions. Moreover, it is obvious that the overall stiffness modulus of the overlay decreases during the cracking process. This can imply that the reinforcing effect of a given product becomes only apparent during the crack propagation phase and not yet in the initiation phase.

Table 3.2 gives an overview of the roles of the different types of interlayer systems.

Table 3.2 Role of interlayer systems.

Function	Reinforcement product	Flexible product that resists at high strains. Controlled debonding	Waterproofing
Sand asphalt		X	X
SAMI		XX	XX
Impregnated nonwoven		XX	XX
Grid	X/XX (*)		X (**)
Steel reinforcing netting	XX	X(***)	X (**)
Three-dimensional steel honeycomb grid	XX		
Grid on nonwoven	X/XX(*)	XX	XX

X : efficient ; XX : highly efficient.
(*) : Reinforcement function depends on product and on temperature conditions.
(**) : Only in case of grids or steel reinforcing nettings embedded in slurry seal or surface dressing.
(***) : Only in case of steel reinforcing nettings embedded in a slurry seal with elastomeric binder.

It is necessary in all cases that there is a sufficient adherence of the interlayer system to the underlayer and to the bituminous overlay in order to guarantee a good distribution of the stresses induced by the traffic over the entire road construction. Poor adherence can result in the rapid development of overall fatigue or in delamination with the appearance of secondary cracks [ref. 1].

In case no waterproofing is guaranteed, an additional waterproofing layer can be added.

3.3.1.3 The bituminous overlay

Apart from the interlayer system itself, the thickness and mix design of the bituminous overlay play important roles for the resistance to reflective cracking of an overlay system.

*a. **Influence** of the thickness of the asphaltic overlay*

Increasing the thickness of the bituminous overlay on a cracked layer is a very effective means to slow down the reappearance of cracks at the pavement surface, since a thicker overlay greatly reduces the stresses induced by traffic at the existing cracks (initiation stage). A thicker overlay also lengthens the path to be followed by an incipient crack before it reaches the surface (propagation stage). As for reflective cracking which is the result of thermal movements of the base layer, a greater thickness offers a protection which mitigates differences in temperature, thus reducing the amplitude of movement of crack edges (initiation stage).

Although the overlay thickness has a strong influence on the time when reflection cracks reappear, it was observed for thick overlays that cracks often initiate at the top of the bituminous layer and propagate downwards to meet the crack in the underlying layer [ref. 9, 10, 11]. In these cases, the onset of reflective cracking is largely determined by the properties of the wearing course.

b. Influence of the mix design of the asphaltic overlay

The resistance of a bituminous material to cracking depends mainly on the nature of the aggregate, the content of bituminous binder, and the characteristics of that binder. The aggregate plays a part through its coefficient of expansion (sensitivity to variations in temperature) and the quality of the bond to the binder (adhesion). These factors may, however, be considered as subsidiary to binder content and characteristics. The latter play a leading part in the material's resistance to cracking, as they directly affect :

- Its reversible elastic deformability - especially at low temperatures -, i.e., its capability of absorbing the stresses transferred at cracks,
- Its self-healing properties, i.e., the possibility for cracks to close and mend under traffic in summer,
- Its resistance to ageing, i.e., the capability of the material of retaining the two above-mentioned characteristics over time.
- The magnitude of thermal stresses which reduce the "stress reserve capacity" of the structure when it is subjected to large temperature variations under cold climatic conditions [ref.12].

To achieve these various properties, a somewhat viscous binder must be selected and binder content must be high. Unfortunately, the freedom of choice between binders is limited by considerations such as rutting and skidding resistance.

Various solutions have been suggested to meet this combination of objectives :

- The use of polymer-bitumens :

  The polymers mainly used are styrene-butadiene-styrenes (SBS) and ethylene-vinyl acetates (EVA). The former behave as elastomers while the latter are plastomers. Admixing them in sufficient amounts to selected bitumens results in binders which are less temperature susceptible and which have markedly higher viscosities at the temperature of working than the pure bitumens normally used. This makes it possible to increase the binder content of the mix.

- The use of bitumens modified with powdered recovered rubber :

  This technology allows to produce very viscous binders which can be used at high contents in bituminous mixes.

- The addition of fibres :

  In this technique, the modification does not affect the binder but the mastic or the mix. The fibres used are very thin and short. Some are mineral in nature (e.g. natural and artificial rock fibres, glass fibres), others are organic (e.g. cellulose fibres) or, exceptionally, are made of steel. Selected fibres have capacities of fixing increased quantities of bitumen and of reinforcing the coating mastic. Adding fibres enables developing mixes rich in bitumen and, therefore, displaying high resistance to moisture, ageing, fatigue, and cracking [ref. 13, 14, 15].

3.4 Conclusions

In this chapter, methods are described to reduce the appearance of reflective cracks at the road surface. A first category of methods prevents or reduces the formation of cracks during the initial construction phase of the road or makes unavoidable cracks less active. Some examples are : the choice of the base materials, the use of a pre-cracking technique during the construction of layers treated with hydraulic binders, the use of crack injection techniques and crack and seat techniques.

The second category of methods consists in the use of an overlay system, meaning the combined system of a bituminous overlay, interlayer system and levelling course, placed on an underlying road structure, usually after prior repairs. One of the components may be absent, depending on the condition of the old road structure, the loading conditions, and on the type of rehabilitation system which is chosen.

It is emphasized that the success of an overlay system depends on the performance of its individual components and on their combined performance as a system. Concerning the overlay itself, the thickness and the mix design play an important role for the resistance to reflective cracking.

For interlayer systems, it is shown that the large variety of commercially available interlayer products, can be classified into a limited number of categories : sand asphalt, SAMIs, nonwovens, grids, steel reinforcing nettings, three-dimensional steel honeycomb grids, and combinations of these products. How to characterize these interlayer systems and how to determine their performance as part of an overlay system is described in the fourth chapter of this book.

Depending on the interlayer product, its role can be :

- reinforcing, to take up the large localised stresses in the vicinity of cracks,
- "controlled debonding", to provide a layer able to deform horizontally without breaking and to allow the large movements taking place in the vicinity of cracks,
- waterproofing.

It is emphasized that a good adherence must be present between the interlayer system and the underlayer and the bituminous overlay. A good functioning also requires a correct placement of the interlayer systems. A detailed description of the procedures used for the different types of interface systems is therefore given in the sixth chapter.

3.5 References

1. M. G.Colombier : "Retarding measures for crack propagation : state of the art". Proceedings of the 2nd International RILEM Conference, Liège, 1993.

2. PIARC Technical Committees on flexible roads and concrete roads, "Semi-rigid pavements", Paris (France), 1991.

3. G. Colombier and J. P. Marchand : "The precracking of pavement underlays incorporating hydraulic binders", Proceedings of the 2nd International RILEM Conference on Reflective Cracking in Pavements", pp. 273-281, 1993.

4. M. Lefort : "Technique for limiting the consequences of shrinkage in hydraulic-binder-treated bases", 3rd International RILEM Conference on Reflective Cracking in Pavements", pp. 3-8, 1996.

5. F. Verhee : "Préfissuration par film plastique", Proceedings of the 3rd International RILEM Conference on Reflective Cracking in Pavements", pp. 47-54, 1996.

6. J. Samanos, H. Tessonneau : "New system for preventing reflective cracking : membrane using reinforcement manufactured on site (MURMOS)", Proceedings of the 2nd RILEM-Conference on Reflective Cracking in Pavements, pp. 307-315, 1993.

7. C.R.O.W. - publication No.69 "Asfaltwapening : zin of onzin ?", 1993.

8. H. Tessonneau and J. Samanos : "Revue Générale des Routes et Aérodromes", No.713, December 1993.
   H. Tessonneau and J.C. Roussel : "Revue Générale des Routes et Aérodromes", No.680, December 1990.

9. M. E. Nunn : "An investigation of reflection cracking in composite pavements in the United Kingdom", Proceedings of the 1st International RILEM-Conference on Reflective Cracking in Pavements, pp. 146-153, 1989.

10. M. E. Nunn and J.F. Potter : "Assessment of methods to prevent reflection cracking", Proceedings of the 2nd International Conference on Reflective Cracking in Pavements, pp. 360-369, 1993.

11. M.D.Foulkes : "Assessment of Asphalt Materials to relieve Reflective Cracking of Highway Surfacings", Ph.D.- Thesis, Plymouth Polytechnic, Plymouth, UK, 1988.

12. W.Arand : "Behaviour of asphalt aggregate mixes at low temperatures". Proceedings of the fourth RILEM symposium on Mechanical tests for bituminous mixes, Budapest, pp 68-84, 1990.

13. J. P. Serfass, J. Samanos : "Bétons bitumineux avec fibres". Revue Générale des Routes et des Aérodromes. Nr 725, pp. 52-60, 1995.

14. J. P. Serfass, B. Mahé de la Villeglé : "Apports des enrobés avec fibres dans la lutte contre les remontées de fissures". Proceedings of the 3rd International RILEM-Conference on Reflective Cracking in Pavements, pp. 199-208, 1996.

15. J. P. Serfass, J. Samanos : "Fiber modifed asphalt concrete characteristics, applications and behaviour. Association of Asphalt Paving Technologists. Volume 65, pp. 193-230, 1996.

4

Characterization of overlay systems

A. Vanelstraete, A.H. de Bondt, L. Courard

In order to choose a suitable overlay system for a given situation, the characteristics which are relevant for the role they have to play in the road structure have to be determined. Limits for these characteristics can then be specified in standard tender specifications. Tests are therefore necessary on the interlayer system itself, on the overlay and on the behaviour of the interlayer as part of the overlay system. The characterization of interlayer systems is described in section 4.1. Tests to evaluate the effect of the interlayer system as part of the overlay system are given in section 4.2. Tests on overlays are not described here, an overview has been given in [ref. 1].

4.1 Characterization of an interlayer system and its components

Characterization of the interlayer system implies that tests are carried out on the interlayer product itself to determine its properties. The effect of the fixing layer/system is another important aspect, which can however only be evaluated fully, if it is considered as part of the interlayer system and overlay system. It is therefore discussed in section 4.2.

For what concerns interlayer product characterization, one has to distinguish between bitumen based interlayer products, such as sand asphalts and SAMIs (see 3.3.1.2), and other types of interlayer products, such as nonwovens, grids and steel reinforcing nettings.

4.1.1 NONWOVENS, GRIDS, STEEL REINFORCING NETTINGS

The properties which are relevant and are often used in standard tender specifications are listed below. Some of these are described in more detail in 4.1.1.1 to 4.1.1.6.

- base material : polyester, polypropylene, fibre glass or steel,
- thickness (in mm),
- rigidity of the junctions (see footnote 2 and fig.3.6 in chapter 3),
- mesh width (this is only relevant for grids and steel reinforcing nettings),
- ultimate strength (in kN/m),
- strain at ultimate strength (in %),
- product stiffness (in kN/m),
- stiffness modulus of the interlayer material (in MPa),
- temperature susceptibility,
- quantity of bitumen that can be absorbed by the interlayer product (this is only relevant for nonwovens).

Prevention of Reflective Cracking in Pavements. Edited by A. Vanelstraete and L. Francken. RILEM Report 18. Published in 1997 by E & FN Spon, 2–6 Boundary Row, London SE1 8HN. ISBN 0 419 22950 7.

4.1.1.1 Thickness

Thickness measurements can be performed according to the EN 964-1 or ASTM D 1777 standards.

4.1.1.2 Mesh width

The mesh width has to be sufficiently large in comparison with the maximum grain size of the overlay, in order to allow the aggregates to penetrate through the mesh. Otherwise, unadequately compacted spots may remain after overlaying or insufficient adherence with the overlay may be the result. In case of using slurry seals, SAMIs or surface dressings with thicknesses larger than the product thickness, aperture size is of less importance, since then overlay and old surface are sticked together by means of this intermediate layer.

4.1.1.3 Ultimate strength and strain at ultimate strength

Simple tensile tests (see fig. 4.1) can be carried out to determine the ultimate strength, strain at failure, product stiffness and stiffness modulus of interface products. Many interlayer products are highly anisotropic and need to be tested, both in longitudinal and transverse direction.

The ISO 10319 standard applies to most nonwovens and also to grids but in this case specimen dimensions may need to be altered. For grids, the ISO 5081 standard is also often used (see further in this section). These test methods cover the measurement of load-elongation characteristics and include procedures for the calculation of stiffness, ultimate strength per unit width and strain at ultimate strength. Singular points in the load - extension curve are also indicated.

In spite of the international standard testing procedures different testing methods are still used in different countries, sometimes leading to differences in the results and making it difficult to compare the characteristics of these products : some of them are given in table 4.1.

Table 4.1 Examples of national standards for tensile tests on nonwovens and grids.

Interlayer product	Standards
Nonwovens	NFG38-014 ; DIN 53857 ASTM D4595-86 ; NBNB 29001 BS 6906
Grids	ASTM D 4595-86 DIN 53857

Different testing devices and conditions are used for interlayer products depending on their base material and their type :

- Grids which are not rigid in the junctions (so-called wovens and nets) are usually characterized by tensile tests on one strand (single end testing method : e.g., ISO 5081 - method) and not on several elements. This is most often the case for fibre glass grids.

- Grids which are completely rigid in the junctions can be characterized with tensile tests on the grid itself. This is the multiple end method (ISO 10319-method).

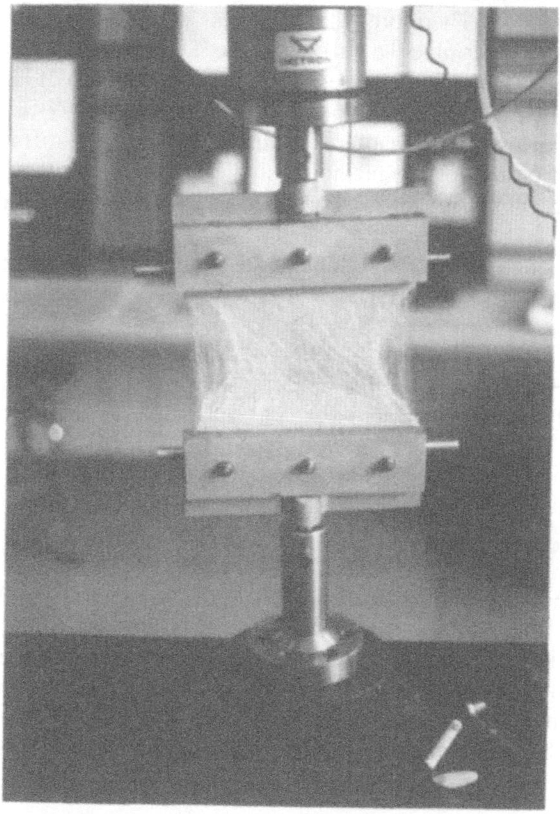

Fig. 4.1 Example of a simple tensile test on a nonwoven (OCW-CRR 3423/19).

- Special precautions and test equipments are often necessary to avoid problems related to clamping. The specimen can either be too loosely fixed, leading to sliding at the fixing system, or can be too firmly fixed, giving rise to crushing at the clamps. The ISO 10319 standard describes different types of jaws for multiple end testing of nonwovens and grids. To avoid breaking or sliding at the clamps in case of multiple end testing, grids can be glued with epoxy in moulds which are then installed in the clamps. The new EN-standard (equivalent to the ISO 10319 standard) will impose compressive block jaws (see ISO 10319 standard).

4.1.1.4 Product stiffness and stiffness modulus

Values of the ultimate strain and strength are not the most important parameters for interlayer systems, since these values are generally not reached in actual road conditions. More important are values of stiffness and stiffness modulus at certain strain-levels (e.g., at 2 % strain).

Interlayer products can in a road structure be considered as two-dimensional foils. The product stiffness S is given by: $S = F/\epsilon$, in which F is the applied force in kN/m in a simple tensile test and ϵ is the corresponding strain. Interlayer products in general do not behave linearly. Therefore, their stiffness depends on the level of force or strain : e.g., one can determine the stiffness, at x % of ultimate strength or at y % of failure strain. ISO 10319 defines the secant stiffness as the ratio of load per unit width to a given value of strain.

In the case of grids, the way of determining the product stiffness and stiffness modulus depends on the testing method : single end or multiple end method. The product stiffness S (in kN/m) for a given level of stress or strain is determined as :

- in case of single end testing methods (testing on one strand) :

  $$S = (F / \epsilon) . n$$

  in which F is the applied force in kN, ϵ is the corresponding strain and n is the number of strands per meter,

- in case of multiple end test methods :

  $$S = F / \epsilon$$

  in which F is the applied force in kN/m and ϵ the corresponding strain.

For steel reinforcing nettings, the stiffness is deduced from that of the used steel wire, taking into account the quantity which is present in the actual steel netting.

The stiffness modulus for nonwovens can be calculated from the stiffness S by : $E = S/d$, in which d is the thickness of the nonwoven. For grids, two definitions can be considered :

- An average or equivalent stiffness modulus : the grid is considered as a continuous foil with a stiffness modulus of $E = S/d$, in which d is the thickness of the grid and S is the product stiffness.
- The stiffness modulus of the strand/rib. This stiffness modulus is calculated as :

  $$E = S . (b/b') /d$$

  in which d is the thickness of the grid and b/b' is the fraction of the total grid occupied by the ribs (see figure 4.2). This value is generally lower than the stiffness modulus derived from that of the base material, because of the use of coatings.

The second definition is most realistic to be used in modelling (see chapter 5).

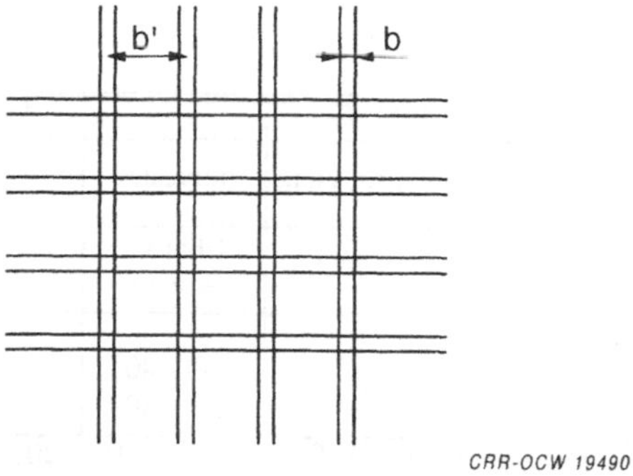

Fig. 4.2 Definition of b and b' for grids.

It is clear from table 4.2 that the characteristics of nonwovens and grids are quite different; this leads to a different role they play in the road structure.

Nonwovens are characterized by a large strain at ultimate strength ; they can withstand large deformations in their plane. Hence, if large horizontal deformations take place at the crack tip, e.g., as a result of temperature variations, they can withstand them, whereas asphalt cannot. In this way they can prevent the crack to propagate upwards. However, their product stiffness and stiffness modulus is much smaller than that of grids, which act as reinforcement. The stiffness modulus indeed indicates whether or not the interlayer product acts as reinforcement product. Its value has to be compared with that of the overlay, which is not only temperature and frequency dependent, but decreases also during the cracking process of the overlay. Since the stiffness modulus of the overlay is highly temperature dependent (above 15000 MPa below 0°C, between 7000 - 15000 MPa between 0 and 15°C, and below 5000 MPa or even 3000 MPa above 15°C, depending on frequency and mix composition), it follows from table 4.2 that some grids can act as reinforcement product in summer, but not in winter conditions. Even so, because of the frequency dependence of the overlay, a given product may be reinforcing at slow loading but not at rapid loading. It is clear that most of the grids have their largest contribution in retarding cracking during the propagation phase.

4.1.1.5 Temperature susceptibility

The determination of the Vicat softening point indicates whether there are high risks for melting or deformations of the product during placement. As can be seen in table 4.1, this is a risk for polypropylene. It is not recommended to use grids made of polypropylene in combination with bituminous overlays for which the temperature during placement exceeds 148°C without special protection precautions (see chapter 6).

Courard et al. [ref. 2] studied the temperature increase of an impregnated nonwoven made of polypropylene after spreading out a bituminous mix of 160°C. It was observed that the temperature in the nonwoven did not pass 80°C, mainly because of the protective effect of the binder layer.

Table 4.2 Indicative values for the basic characteristics of commercially available nonwovens and grids.

	Nonwoven		Grid		
Base material	polyester	polypropylene	polyester	polypro-pylene	fibre glass
Type	-	-	fillet or woven	grid	fillet or woven
Mesh size (mm) Longitudinal direction Transverse direction	-	-	 20 - 40 20 - 40	 ~ 50 ~ 75	 10 - 40 10 - 40
Mass per unit area (g/m²)	130-160	130-160	200 - 500	200 - 300	200 - 650
Thickness (mm)	0.7 - 1.6	0.7 - 1.6	0.7 - 1.2	0.7 - 1.1	0.7 - 1.2
Ultimate strength (kN/m) Longitudinal direction Transverse direction	 5 - 10 5 - 10	 5 - 10 5 - 10	 50 - 90 (1) 50 - 90 (1)	 14 - 25 (2) 18 - 25 (2)	 35 - 100 (3) 50 - 200 (3)
Strain at ultimate strength (%) Longitudinal direction Transverse direction	 35 -100 35 -100	 50 - 90 50 - 90	 10 - 15 (1) 10 - 15 (1)	 10 - 15 (2) 8 - 10 (2)	 3 - 4 3 - 4
Product stiffness at 2 % strain (kN/m)	10 - 20	10 - 20	200 - 400	400 - 800	1500- 8000
Stiffness modulus at 2 % strain (MPa) (4)	7 - 30	7 - 30	3000 - 6000	4000 - 5000	4000 - 7000
Vicat softening point (°C)	230 - 240	148	230 - 240	148	> 300
Quantity of absorbable binder (kg/m²)	0.7 - 1.4	0.7 - 1.4	-	-	-

(1) Determined from single end tensile tests (ISO 5081 - method).
(2) Determined from multiple end tensile tests. The values may vary depending on the used method.
(3) Determined from single end tensile tests with capstan grips.
(4) According to the second definition given above for the stiffness modulus of grids. This value is generally lower than the stiffness modulus derived from that of the base material, because of the use of coatings.

4.1.1.6 Absorbable quantity of bitumen

For nonwovens, laboratory tests are necessary to determine the quantity of bitumen they can absorb. An insufficient quantity of binder indeed leads to improper functioning of the interlayer due to delamination by poor bonding. An excess of binder can lead to problems during placement due to sticking to the vehicle tyres with possible detachment of the interlayer product from the underlayer. Specimens of the nonwoven are therefore impregnated with binder and then leaked out during well-determined periods of time. The quantity of absorbed bitumen is simply determined from the difference in the specimen's weight before and after impregnation [ref. 3,4].

4.1.2 BITUMEN BASED INTERLAYER PRODUCTS

The characterization of sand asphalt is similar as for bituminous overlays, e.g., determination of stiffness moduli, fatigue behaviour, resistance to thermal cracking and resistance to permanent deformation. These properties are strongly temperature and frequency dependent. They highly depend on the rate of bitumen and on the void content of the mix.

For SAMI's, the stiffness modulus can be determined from rheological measurements. Tensile strength and strain depend highly on temperature and loading time.

Bitumen based interlayer products are most often prepared with polymer modified binders. This improves their performance, e.g. to withstand large deformations in their plane.

4.2 Characterization of overlay systems

In addition to the tests described in 4.1 to determine the basic characteristics of interlayer products, it is also necessary to study the behaviour of interlayer systems as part of the overlay system. Laboratory tests to study the effect of interlayer systems as part of an overlay system concern mainly two types of tests :

- Tests to determine the adherence of the interlayer system with the underlayer and asphaltic overlay, in the plane of the interlayer and perpendicular to it. The adherence will depend on the fixing method/layer, on its quantity, and also on the type of interlayer product.
- Laboratory tests to investigate the performance of overlay systems. Here, the effect of a given interlayer system may depend on the exact loading conditions. Therefore different loads must be considered : repeated thermal loading, traffic loading (with or without large vertical displacement at the crack edges). Some products may be very efficient in one case, but totally ineffective in the other.

4.2.1 ADHERENCE TESTS ON OVERLAY SYSTEMS

Simple tensile tests, pullout tests and shear tests are generally used to study the adherence between the interlayer system and its surroundings. They are briefly discussed below.

4.2.1.1 Simple tensile tests

Simple tensile tests, schematically represented in figure 4.3, can be performed to investigate the adherence of the interlayer system with the underlayer and the bituminous overlay, most often due to actions performed perpendicular to its plane. The results depend on the testing conditions such as the temperature and deformation rate.

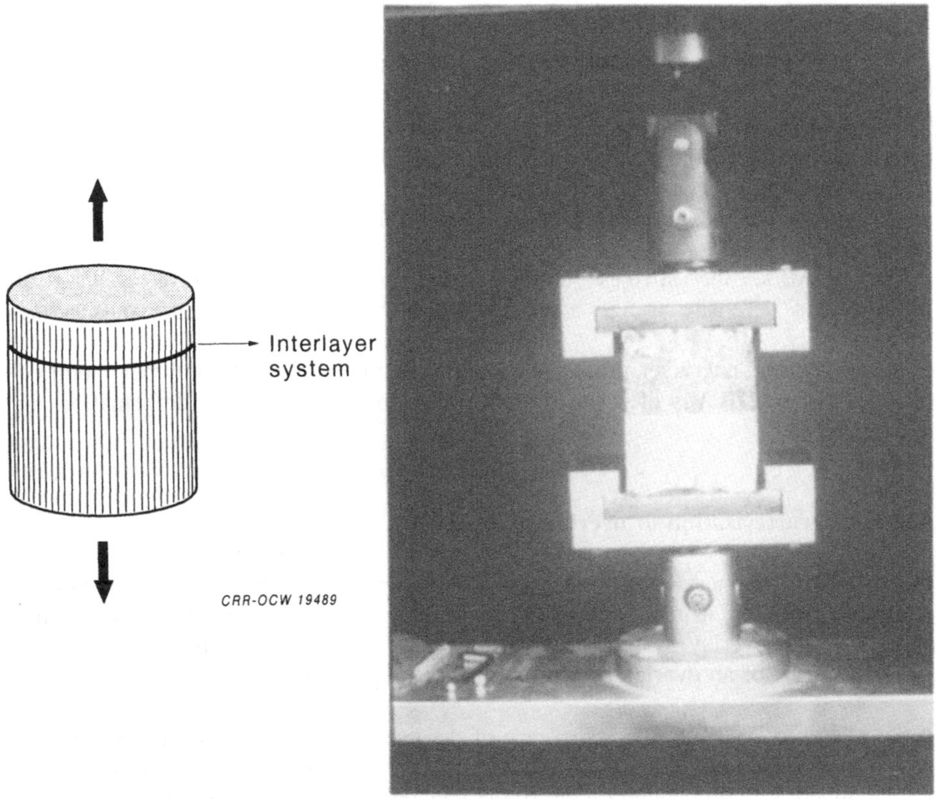

Fig. 4.3 Schematic representation of a simple tensile test.

4.2.1.2 Pullout tests

To get information on the pullout behaviour of interlayer systems, so called pullout tests can be performed (see figure 4.4). A pullout force (external load) versus pullout slip relationship is obtained. Effects of individual characteristics of the interlayer system (such as product stiffness, type and amount of adherence product) can only be found by means of parametric testing.

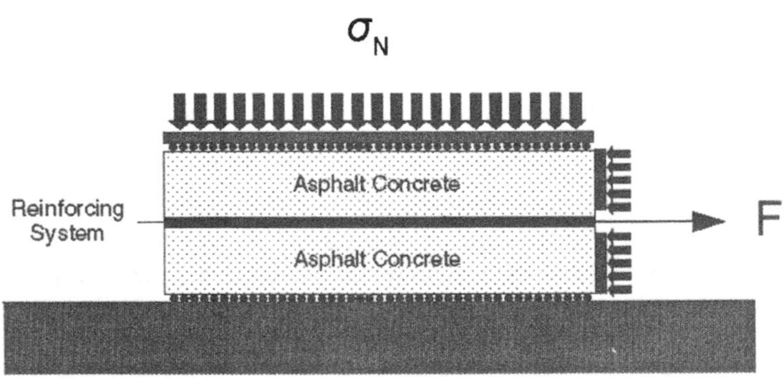

Fig. 4.4 Sketch of pullout test.

4.2.1.3 Shear tests

- Direct shear tests
  Information on the shear response of the interaction between interlayer product on one hand and underlayer and overlay on the other hand can be determined via the direct shear test set-up (see figure 4.5) which has been developed by Leutner [ref. 5]. Cores with a diameter of 150 mm can, without any preparations, be used directly in the testing process. The "output" of these tests are shear stress versus shear slip relationships, which can serve as input for finite element packages. Figure 4.6 shows an example of curves obtained from Leutner's test for four specimens [ref. 6]. It can be observed that although cores were drilled next to each other from the same paving track (see inset), quite a large variability occurs. Four specimens per tested situation seems therefore to be a minimum allowable number.

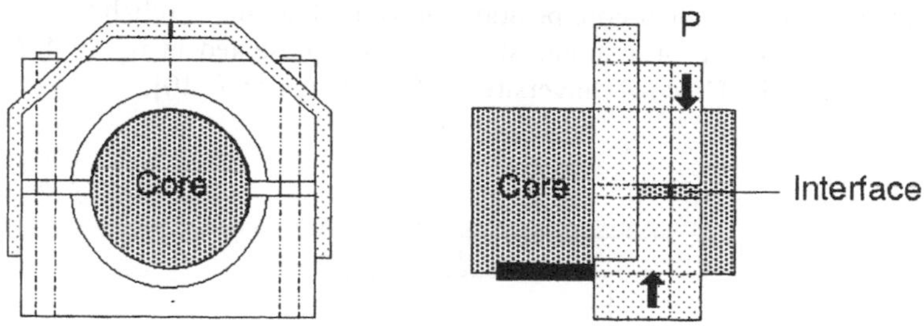

Fig. 4.5 Sketch of Leutner's Direct Shear Test Set-Up [ref. 5].

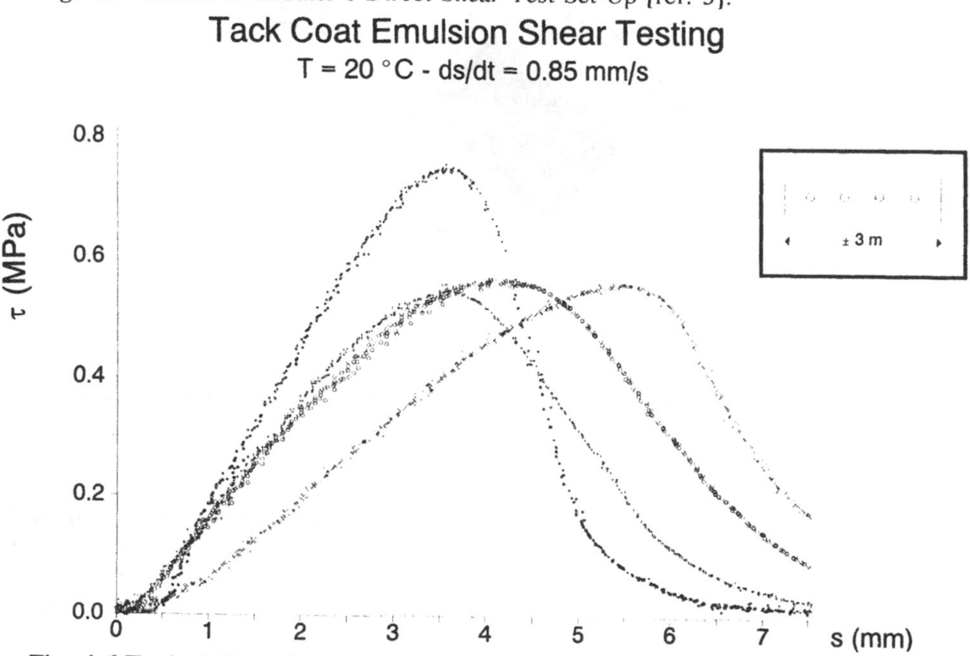

Fig. 4.6 Typical Shear Stress versus Shear Slip Relationships [ref. 6].

It must not be forgotten that although the tests mentioned above are attractive, because of their simplicity, they still show the drawback of the family of direct shear tests set-ups : the presence of a bending moment at the shear plane. Linear elastic finite element analysis [ref. 7, 8] revealed that for the set-up of Leutner, a) the shear stress distribution along the shear plane is uniform in case of soft adhesion (or bond) and non-uniform in case of stiff adhesion (or bond) and b) the normal stress distribution along the shear plane is non-uniform. From this analysis it also became apparent that the shear stress should be computed by dividing the shear force by the cross-sectional area of the core; this is because the slender beam theory is not valid anymore. However, because of the fact that routine laboratories equipped with a Marshall press only, are still capable of carrying out these tests, it has been concluded that for every day designs, it is a well suited tool.

Modifications of Leutner's Shear Test were made by Delft University of Technology [ref. 9]. The modifications allowed that normal pressure could be applied onto the shear plane and enabled core positioning to be done more reliably.

Another example of a simple shear test is represented in figure 4.7 and has been developed at the Cracow University of Technology [ref. 10].

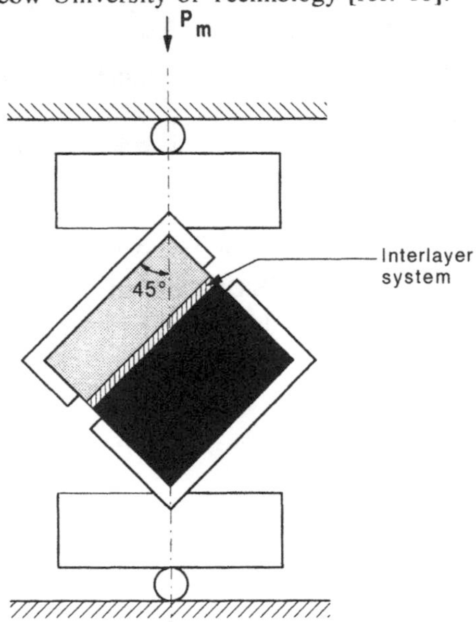

Fig. 4.7 Simple shear test developed at the Cracow University of Technology [ref. 10].

- 4-Point Shear Tests

  An example of a 4-point shear test is schematically represented in figure 4.8 [ref. 7,8]. Due to the position of the supports no bending moment is generated along the shear plane; it is thus a pure shear test, which shows uniform shear and normal stress distributions. However, due to its complexity it is only suited for research purposes. It has been applied in the past to shear unreinforced as well as reinforced cracks [ref. 6].

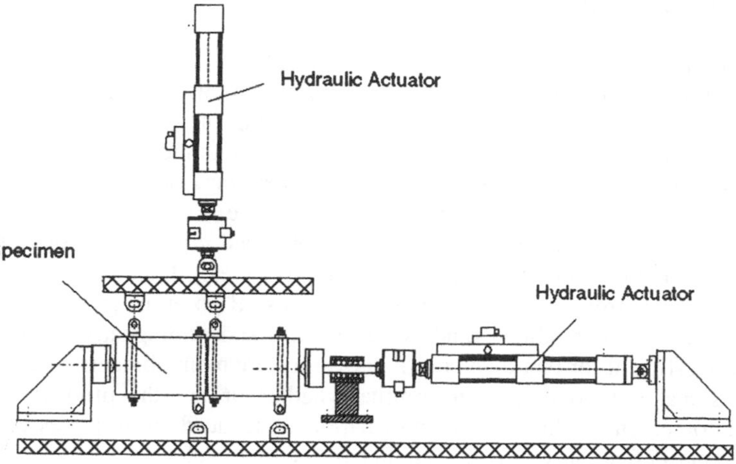

Fig. 4.8 Example of a 4-Point shear test [ref. 8].

4.2.2 LABORATORY TESTS FOR OVERLAY SYSTEMS UNDER REPEATED THERMAL AND TRAFFIC LOADING

Several test devices exist to investigate the efficiency of overlay systems under thermal, traffic or combined thermal and traffic loading. In all tests the onset of a crack and its progression in function of time or number of loading cycles are monitored. Table 4.2 gives an overview of the existing testing facilities. The tests are generally carried out for comparative purposes, however, in some cases they are used to calibrate and to verify computational models. A brief description of the different testing facilities is given below.

Table 4.2 Overview of different testing procedures.

Type of loading	Testing procedure	Examples of references
Traffic loads	- Beam bending testing facility - Wheel tracking testing equipment - Simulation test of rocking overlaid cement concrete slabs	11, 12, 13 12, 14 15
Thermal loads	- Test for thermal shrinkage of base layer - Test for thermal shrinkage of upper layer	16, 17, 18 19, 20, 21, 22
Traffic and thermal loads	- Combination of beam bending test facility and thermal shrinkage of base layer	23, 24

4.2.2.1 Simulation of traffic loads

- Beam bending testing facility [ref.11, 12, 13]
  The test structure, consisting of an overlay system on a cracked subbase, is cut under the form of a beam. The beam is resting on a flexible support or on rolling contact points. Repeated loads are applied at the surface of the overlay in order to simulate either mode 1 (pure bending) or mode 2 (pure shear) loading conditions. An example of the test device is represented in figure 4.9. As a result of these tests, curves can be obtained giving the evolution of the crack length versus the number of loading cycles. Examples are shown in figure 4.10 [ref.13]. From this type of tests, Rigo et al. [ref.13] concluded that the use of modified binders for SAMIs and impregnated nonwovens is more efficient for retarding cracks in overlays than pure binders. Beam testing tests make it possible to determine whether or not the product acts as a reinforcement. This type of information is deduced from curves giving the vertical deflection versus the number of loading cycles.
  The most critical problem of the beam bending test appears at the boundaries of the sample where the reaction forces of the rest of the road structure have to be introduced, usually in a very approximate manner.

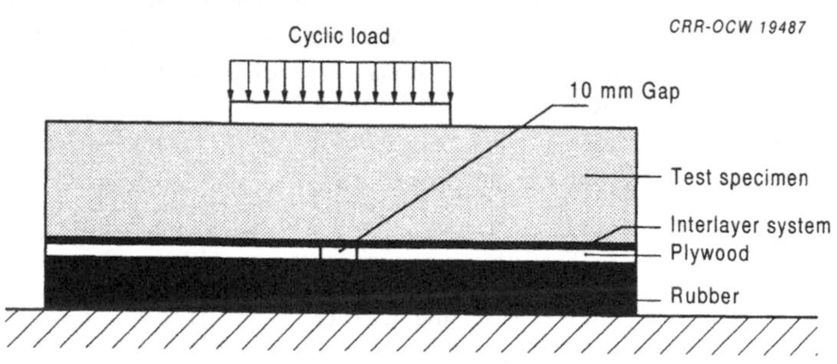

Fig. 4.9 Schematic representation of the beam bending testing facility of [ref.12].

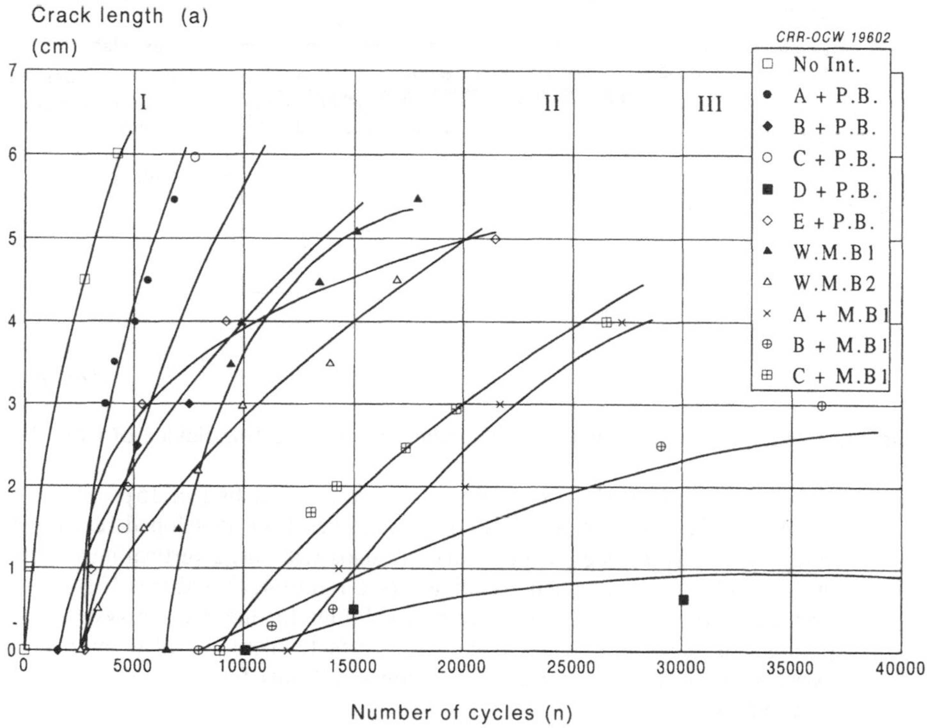

No Int.	: without any interlayer system
W.M.B1	: with a layer of modified bitumen No. 1 (no nonwoven)
W.M.B2	: with a layer of modified bitumen No. 2 (no nonwoven)
A, B, C, D, E	: five types of nonwovens
A, B, C, D, E + P.B.	: nonwoven A, B, C, D, E + pure bitumen
A, B, C + M.B1	: nonwoven A, B, C + modified bitumen No. 1

Fig. 4.10 Example of results from beam bending tests obtained in [ref. 13] on nonwovens and SAMIs.

- Wheel tracking testing equipment [ref. 12, 14]
  A slab or beam representing the test structure is submitted to the action of a moving wheel. It allows investigation of reflective cracking on larger samples than permitted by the beam testing method, and under more representative loading conditions. A schematic representation of the wheel tracking testing equipment of [ref.12] is given in figure 4.11. Modelling of boundaries seems to be a weak point, also in these cases.

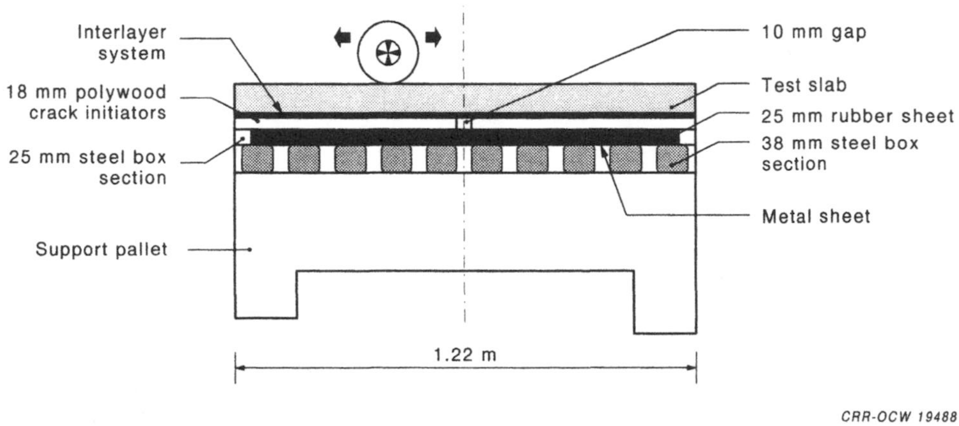

Fig. 4.11 Schematic representation of the wheel tracking testing device of [ref.12].

- Simulation test of rocking overlaid cement concrete slabs [ref.15]
  At the Belgian Road Research Centre a test has been developed recently to study overlay systems on cement concrete slabs with large vertical movements at the crack edges, so-called "rocking" (see figure 4.12 and [ref.15]). These vertical movements imply very severe conditions for the overlay system. The purpose of the study is to compare the effectiveness of different overlay systems and to determine maximum allowable limits for the relative vertical movements.

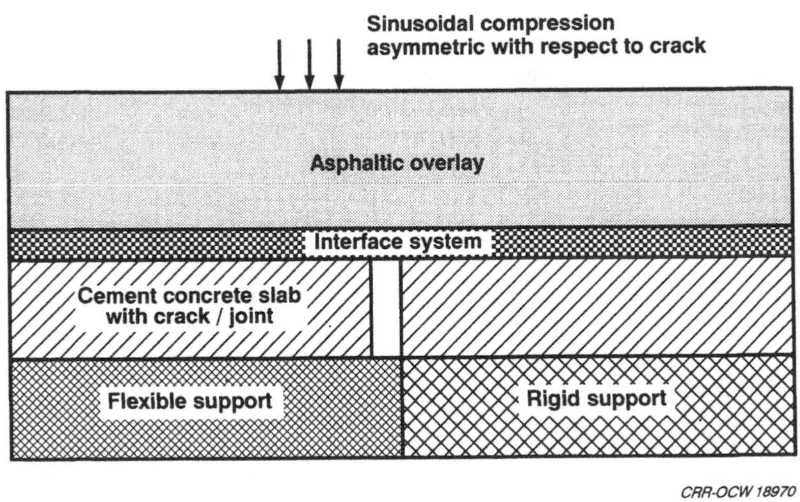

Fig. 4.12 Schematic representation of the test device of [ref.15] for simulating rocking of overlaid concrete slabs.

4.2.2.2 Simulation of thermal loading

Large strains can develop in base layers as a result of thermal effects, causing joints or cracks in cement based layers to open and close, generally over longer time scales than for traffic loading.

- Laboratory test on thermal shrinkage of the base layer [ref.16, 17, 18].
  Thermal movements of opening and closing of a crack in the base layer can be simulated in different ways, depending on the type of device. In the test device of [ref.16, 17] (figure 4.13), the sample is alternatively submitted to opening and closing cycles of the crack in the base layer. A comparison of the efficiency in crack prevention between different interlayer systems is given in figure 4.14. More details can be found in [ref. 15]. In other equipments (figure 4.15 and [ref.18]) the crack in the base layer is opened continuously at a constant rate.
  The information provided by these tests can vary from simple visual observation of the cracking process in function of loading cycles (or time) to measurements of more fundamental properties such as shear modulus of the interlayer and stress-strain relationships. The tests are generally carried out at low temperatures (below -5°C in most cases) where the risk for cracking in the overlay is the highest.

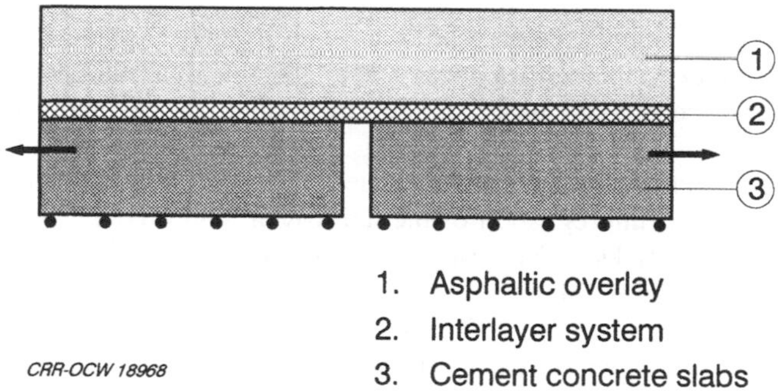

Fig. 4.13 Schematic representation of the BRRC-thermal cracking test [ref. 16, 17].

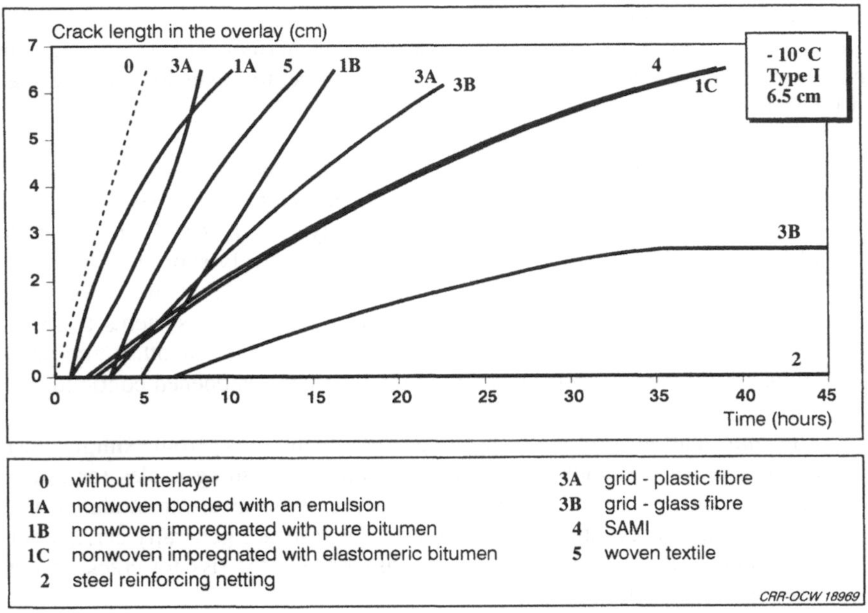

Fig. 4.14 Efficiency of different interlayer systems for the prevention of reflective cracking which is the result of thermal shrinkage of the base layer [ref. 15] : onset and evolution of the crack in the overlay during the experiment. For some types of interlayer systems two extreme curves are given : all tested products belonging to that category lay between them.

As shown in fig.4.14, some systems merely have an influence on the initiation phase of the crack (the crack in the overlay appears at a later time interval) ; others lead also to changes in the propagation phase of the crack (their curves show a decreasing slope).

The results for nonwovens are highly influenced by the type and quantity of the used binder. Only full impregnation of the nonwoven with binder is really efficient for the prevention of cracks. The best results are obtained if the nonwoven is fully impregnated with modified binder. At the testing temperature of -10 °C certain modified binders can become too stiff and crack. Similar conclusions are valid for SAMIs. These products generally perform better at higher testing temperatures (e.g. -5 °C).

The results for grids depend on their type and base material. Better results were generally obtained with fibre glass than with plastic fibre grids. No systematic difference in the results could be observed between grids made of polypropylene or polyester. Evidence was found that the grid size needs to be sufficiently large to allow the stones of the overlay to penetrate through the meshes of the grid. This enables the development of pullout restraint through material being enclosed in the apertures.

The best results in these thermal cracking tests at -10 °C were obtained with steel reinforcing nettings : no crack in the overlay appeared during the experiment.

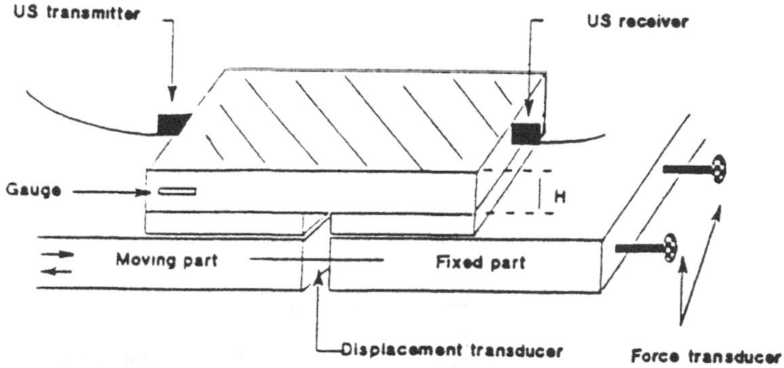

Fig. 4.15 Schematic representation of the ENTPE-apparatus for the simulation of thermal cracking [ref. 18].

- Tests simulating thermal shrinkage of the overlay [ref. 19, 20, 21, 22]. These tests are intended for the simulation of cracks in the overlay which are the result of large temperature variation rates. They give information about the thermal stresses induced in the overlay and the thermal resistance of the overlay. Although this type of test is not directly representative for reflective cracking, the evaluation of the overlay resistance to this type of cracking is an important information. The presence of thermal stresses in the overlay indeed decreases the range of acceptable stresses resulting from other loads [ref. 22]. The test can be carried out on a beam of overlay material, maintained at a constant length in a rigid frame, while changing the surrounding temperature at a constant rate [ref. 19, 20, 21, 22].

4.2.2.3 Traffic loading combined with thermal shrinkage of the base layer [ref. 23, 24]
In order to stick closer with actual conditions dealing with the combined effects of thermal shrinkage and traffic loads, the Public Road Laboratory of Autun (France) developed a testing facility combining the beam bending test with thermal shrinkage at constant displacement rate of the base layer [ref. 23, 24]. The test device of Autun is schematically represented in figure 4.16. A comparison of the efficiency of different interlayer systems is given in figure 4.17. Related to this test, the following recommendations for the comparison of the efficiency of interface systems are used in France :

- ineffective product : time for failure of the specimen in the test < 320 min,
- medium effective : time for failure of the specimen in the test > 320 min and < 415 min,
- effective : time for failure of the specimen > 415 min.

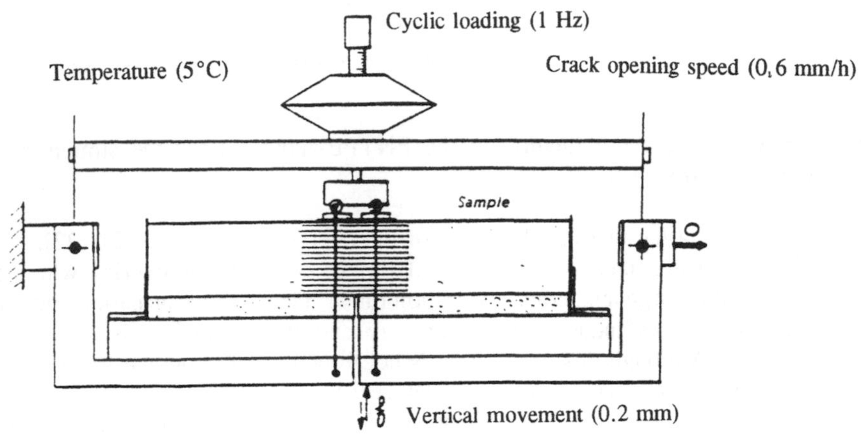

Fig. 4.16 Schematic representation of the test device used at the Public Road Laboratory of Autun (France).[ref. 23, 24].

	Time for crack initiation (min)	Crack growth speed in the 1st cm (μm/min)	Time of propagation (min)	Time for failure (60 mm overlay) (min)
Non-wovens	190 85 - 365	80 30 - 180	550 405 - 730	550 410 - 740
Sand asphalts	105 25 - 175	65 40 - 100	525 380 - 660	530 390 - 665
SAMI's	210 75 - 300	65 20 - 175	530 255 - 730	665 250 - 735

190 → mean
85 - 365 → minimum - maximum

CRR-OCW 19492

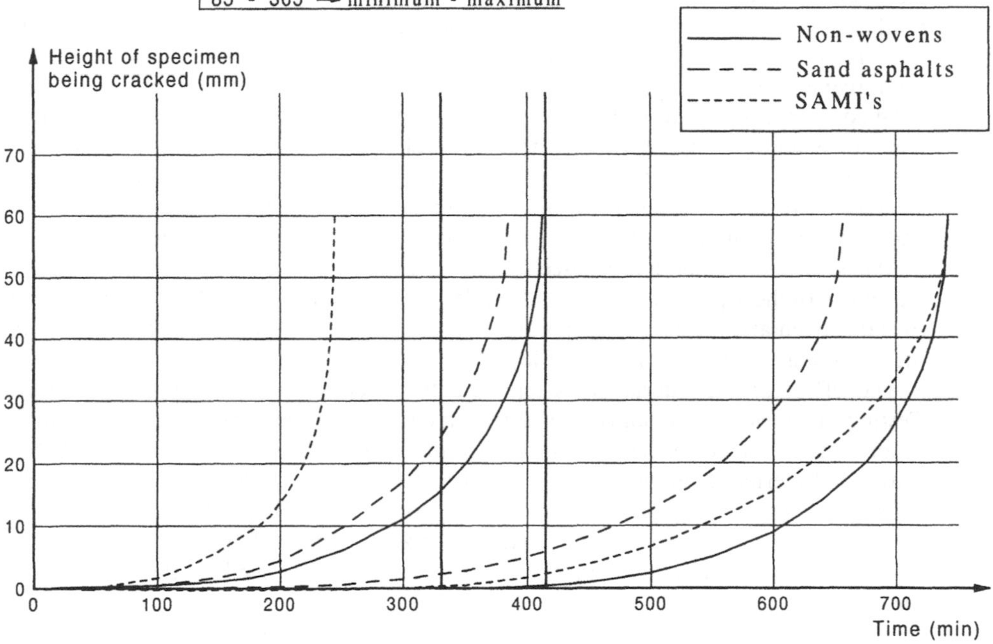

Fig. 4.17 Efficiency ranges of interlayer systems for prevention of reflective cracking in the case of traffic, combined with thermal shrinkage of the base layer. [ref. 23, 24].

4.3 Conclusions

Characterization of interlayer systems implies that tests are performed on the interlayer product as well as on the interlayer system as part of the overlay system. Simple tensile tests are generally performed on the interlayer products to deduce the basic mechanical properties of these products, such as ultimate strength, strain at ultimate strength, stiffness and stiffness modulus. These values are often anisotropic. Different testing methods exist in the different countries and the testing methods also depend on the type of products. Hence, even for two types of grids, the specified values are not always comparable.

From the mechanical properties deduced from tensile tests, the role that interlayer systems play in the road can be deduced. Nonwovens are characterized by a low stiffness modulus, compared to that of the bituminous mix, and are therefore not relevant as reinforcement products. However, they have a high strain at ultimate strength, which makes them suitable to withstand large horizontal deformations as there exist just above the crack tip, e.g., in the case of temperature variations. Hence, they can slow down the crack reflection process in such cases. Grids and steel reinforcing nettings are characterized by a high stiffness modulus. They act as reinforcement product. In order to determine whether or not a given product acts as reinforcement in a given situation, the stiffness modulus of the interlayer product has to be compared with that of the overlay. Taking into account that the overlay stiffness modulus is highly temperature dependent, and changes also with frequency and with the lifetime of the overlay, a given interlayer system can be reinforcing in one situation, and not in another situation (e.g. summer versus winter conditions; initiation versus propagation phase of cracking). Another important characteristic of interlayer products is their temperature susceptibility ; they may not deform or melt during placement of the asphaltic overlay. Also relevant is the quantity of absorbable bitumen for nonwoven and the size for grids and steel reinforcing nettings in comparison with the maximum grain size of the overlay.

The effect of the interlayer system largely depends on the loading conditions. Testing facilities exist to study their effect in case of thermal loading, traffic loading (with presence or not of large vertical movements at the crack edges) and combined traffic and thermal loading. From these laboratory tests, it was found that efficient crack retarding systems do exist, however, not in all loading situations. A lot of interlayer systems are effective in case of horizontal (mode I) crack opening. For nonwovens and SAMIs, their performance is mainly determined by the amount and type of the binder they contain.

With the available test equipments, a qualitative ranking of interlayer systems, under well-defined test conditions, is now possible. Although there seems to be a general agreement between laboratory results and field experiments, most of the laboratory results have however almost never been adjusted to full scale results. And there are even less cases where this adjustment is made on the basis of a statistical analysis from several test sites. At this moment, it is very difficult to predict the improvement in service lifetime of a given interlayer system on the field. Laboratory tests are usually performed under ideal laying conditions, whereas it has been observed worldwide that a lot of projects already failed in a very early stage because of bad placement.

Progress still has to be made in giving recommendations for the laying of the different interlayers systems. This is the subject discussed in the sixth chapter.

4.4 References

1. L. Francken : "Laboratory simulation and modelling of overlay systems", Proceedings of the 2nd International RILEM-Conference on Reflective Cracking in Pavements, keynote paper, pp. 75-99, 1993.

2. L. Courard, J.M. Rigo, R. Degeimbre and J. Wiertz : "Compatibility between fibres and modified bitumen". 1st RILEM-Conference on Reflective Cracking in Pavements, Liège, pp. 103-111, 1989.

3. American Task Force 25 test method "Asphalt Retention and area change of paving engineering fabrics".

4. "Mesure de la quantité de liant retenu dans une membrane géotextile non-tissée pour interlayer anti-fissure", Belgian Road Research Centre, 1996.

5. R.L. Leutner : "Research on Adhesion between layers of Flexible Pavements", Bitumen 3, 1979.

6. A. H. de Bondt : "Anti-reflective Cracking Design of (Reinforced) Asphaltic Overlays", Ph. D.-Thesis, Delft University of Technology, 1997.

7. A.H. de Bondt and A.Scarpas : "Shear Interface Test Set-Ups", Report 7-93-203-12, Road and Railroad Research Laboratory, Delft University of Technology, 1993.

8. A. H. de Bondt and A. Scarpas : "Theoretical Analysis of Shear Interface Test Set-Ups", Report 7-94-203-15, Road and Railroad Research Laboratory, Delft University of Technology, 1994.

9. J.E.F. Berends, A. H. de Bondt, A.Scarpas : "Influence of Bond on the Behaviour of (Un-)Reinforced Asphalt Concrete Overlays", Report 7-94-203-18, Road and Railroad Research Laboratory, Delft University of Technology, 1994.

10. W. Grzybowska, J. Wojtowcz and L.Fonferko : "Application of geosynthetics to overlays in Cracow region of Poland". Proceedings of the 2nd RILEM-Conference on Reflective Cracking in Pavements, Liège, pp. 290-298, 1993.

11 D.Sicard : "Remontée des fissures dans les chaussées. Essais de comportement en laboratoire par flexion sur barreaux", 1st RILEM-Conference on Reflective cracking in Pavements, Liège, pp 71-78, 1989.

12. S.F.Brown, J.M. Brunton and R.J.Armitage : " Grid reinforced overlays", 1st RILEM -Conference on Reflective Cracking in Pavements, Liège, pp. 63-70, 1989.

13. J.M.Rigo et al : "Laboratory testing and design method for reflective cracking interlayers", 1st RILEM-Conference on Reflective Cracking in Pavements, Liège, pp. 79-87, 1989.

14. I.Yamaoka, D.Yamamoto, T.Hara : "Laboratory fatigue testing of asphalt concrete pavements containing fabric interlayers and field", Proceedings of the 1st RILEM-Conference on Reflective Cracking in Pavements, Liège, pp.49-56, 1989.

15. A.Vanelstraete and L.Francken : "Laboratory testing and numerical modelling of overlay systems on cement concrete slabs", Proceedings of the 3rd RILEM-Conference on Reflective Cracking in Pavements, Maastricht, pp. 211-220, 1996.

16. C.Clauwaert and L.Francken : "Etude et Observation de la fissuration réflective au Centre de Recherches Routières Belge", Proceedings of the 1st RILEM-Conference on Reflective Cracking in Pavements, Liège, pp. 170-181, 1989.

17. L.Francken and A.Vanelstraete : "On the thermorheological properties of interlayer systems", Proceedings of the 2nd RILEM-Conference on Reflective Cracking In Pavements, Liège, pp. 206-219, 1993.

18. H.Di Benedetto, J.Neji, J.P.Antoine and M.Pasquier : "Apparatus for laboratory study of cracking resistance", Proceedings of the 2nd RILEM-Conference on Reflective Cracking in Pavements, Liège, pp.179 - 186, 1993.

19. R.A.Jimenez,G.R.Morris and D.A.Dadeppo : "Tests for strain-attenuating asphaltic materials". Proceedings A.A.P.T., vol 48, pp. 163-191.

20. J.Eisenman, U.Lampe, U.Neumann : "Effect of polymer modified bitumen on rutting and cold cracking performance", Proceedings of the 7th Conference on Asphalt Pavements, Nottingham, Vol.2, pp. 83 - 94, 1992.

21. H.Kanerva : "Effect of asphalt properties on low temperature cracking of asphalt mixtures", Proceedings of the 7th Conference on asphalt Pavements, Nottingham, Vol. 2, pp. 95-97, 1992.

22. W. Arand : "Behaviour of asphalt aggregate mixes at low temperatures". Proceedings of the 4th RILEM-Symposium on Mechanical tests of bituminous mixes. Budapest, pp. 68-84, 1990.

23. J.H. Vecoven : "Methode d'etude de systèmes limitant la remontée de fissures dans les chaussées", Proceedings of the 1st RILEM-Conference on Reflective Cracking in Pavements, pp. 57-62, 1989.

24. P.Dumas, J.Vecoven : "Processes reducing reflective cracking : synthesis of laboratory tests", Proceedings of the 2nd RILEM-Conference on Reflective Cracking in Pavements, Liège, pp. 246-253, 1993.

5

Modelling and structural design of overlay systems

L. Francken, A. Vanelstraete, A. H. de Bondt

5.1 Introduction

Reflective cracking is a major concern for engineers facing the problem of road maintenance and rehabilitation. There is evidence after several years of research and practice that materials and procedures having potential to improve the situation do really exist. But it must be emphasized that there is no standard solution suited for every situation.

Assessment has clearly revealed the complexity of the problem and also the fact that the term "reflective cracking" covers a wide variety of phenomena. The success of innovative solutions depends on the correct choice of all the components of an overlay, on their combination and on their implementation in function of the loading conditions to which they will be exposed for a future design life. Typical questions to be answered by road designers are : which surfacing thickness, what type of interlayer system are needed in a given situation ? What is the relative lifetime of a given solution in comparison with a classical solution ?

The present chapter gives an overview of currently used models for the structural design of overlay systems, subjected to temperature variations as well as traffic. Common to all models is that they are constituted by a computing technique to determine the stress-strain distribution in the structure and a set of physical deterioration laws describing the behaviour of the structure and its evolution under service conditions.

5.2 Input data for modelling

Each model requires input data describing the road structure, the loading conditions and the characteristics of the materials.

5.2.1 ENVIRONMENTAL AND LOADING CONDITIONS

The choice of a model is critically dependent on a correct evaluation of the case under study. An extensive review of the possible cases of reflective cracking is given in the first chapter and in [ref. 1].

Prevention of Reflective Cracking in Pavements. Edited by A. Vanelstraete and L. Francken. RILEM Report 18.
Published in 1997 by E & FN Spon, 2–6 Boundary Row, London SE1 8HN. ISBN 0 419 22950 7.

Information is required concerning the road structure :

- The type of road structure (rigid pavement, semi-rigid pavement, flexible pavement).
- The thickness of the layers.
- The type of the discontinuities (road widening, local repairs, transversal or longitudinal trenches, local weaknesses of the subgrade, etc).

Cracks, joints or other types of discontinuities give rise to local stress concentrations. A typical situation is represented in fig. 5.1.

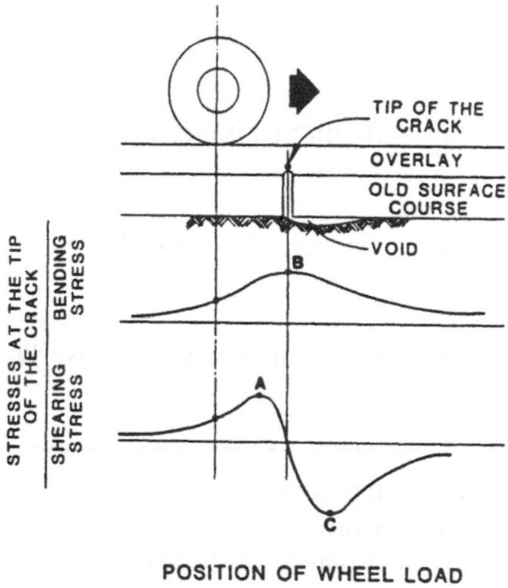

Fig. 5.1 Stresses induced at the cracked section of an overlay due to a moving wheel load [ref. 2].

Concerning the loading conditions, the driving forces for crack initiation and propagation are :

- traffic,
- temperature variations,
- the hydric variations of the soil,
- moisture.

Traffic loads travelling over a crack in a lower base layer generate three successive stress pulses, two in the shearing mode (fig. 5.2a and 5.2c) and one in the opening mode when the load is over the crack (fig. 5.2b).

Temperature changes cause the base layer and overlay to try to expand and contract, generating the mode-I opening of the crack (fig. 5.2d).

In most models these loading types are treated separately in a first approach and superposed afterwards to get a final picture of reality.

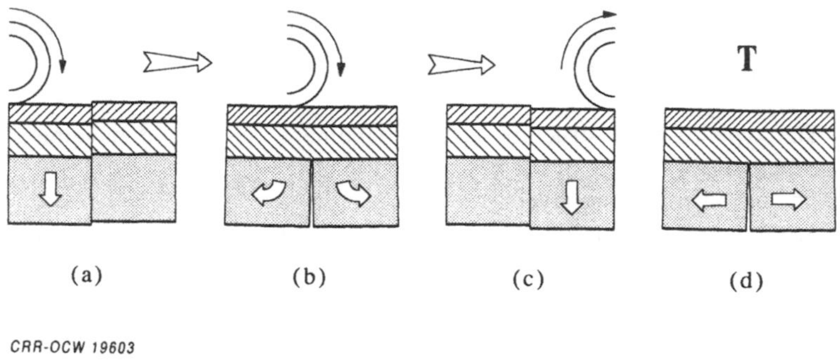

Fig. 5.2 Crack pavement loadings.

5.2.2 CHARACTERISTICS OF THE BASIC COMPONENTS OF AN OVERLAY SYSTEM

The characteristics of the basic components of an overlay system needed for modelling are given in table 5.1.

Table 5.1 Characteristics of the different components of an overlay system needed for modelling.

Bituminous layers	Stiffness modulus Poisson 's ratio Thermal expansion coefficient Fatigue parameters Crack propagation parameters
Fixing layers : binders and tack coat	Penetration, softening point, viscosity, etc ... Shear modulus Temperature susceptibility
Interlayer systems	Product stiffness Stiffness modulus Ultimate strength Strain at ultimate strength Thickness Temperature susceptibility

5.2.2.1 Bituminous overlay materials.
The behaviour of this material in traditional types of overlays is the first information we need in order to have a reference to assess the future performance of more complicated structures. Any evaluation of additional functional layers will indeed be made in comparison with the traditional solution.

Of primary importance is that we have to deal with a material displaying wide variations with temperature and loading time in all its mechanical properties.

The stiffness modulus of bituminous materials

The modulus of bituminous mixes is a complex entity represented by its magnitude, E*, the so-called stiffness modulus and its phase angle, φ. It depends on temperature and loading time. The stiffness modulus can be represented under the form of a master curve (see fig. 5.3) built up from measurements carried out at different combinations of temperatures and frequencies by using shifting factors, log α_T, of the form :

$$\log \alpha_T = \frac{\Delta H}{R} \cdot \left(\frac{1}{T} - \frac{1}{T_s}\right)$$

in which R is the universal gas constant and ΔH the activation energy, $\Delta H \simeq 2 \times 10^5$ J/mol. T and T_s are respectively the test temperature and the reference temperature expressed in K (Kelvin). This modulus master curve can also be estimated from the composition of the material and the binder characteristics [ref. 3, 4, 5].

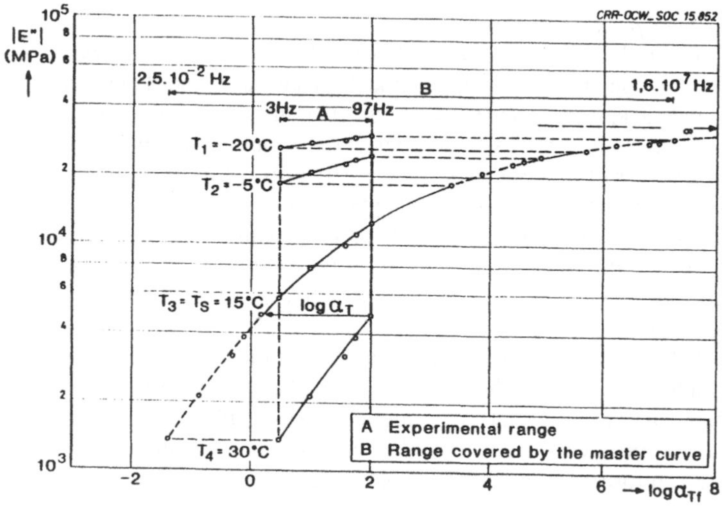

Fig. 5.3 Modulus master curve of a bituminous concrete.

Thermal properties of bituminous materials

It is well known that thermal stresses can induce spontaneous surface cracking under severe winter conditions. Although this effect is not directly relevant for reflective cracking, the presence of such stresses superposed to the other loading conditions can have a strong influence on the resistance of a pavement structure to cracks generated in the bottom layers [ref. 6].

The magnitude of these stresses can be estimated from the thermal expansion coefficient and the stiffness modulus master curve. The resistance of the material to tensile stresses at low temperature can be determined either by tensile tests at low loading rates or by cooling tests on samples maintained at constant length. The critical temperature at which brittle fracture occurs may be determined in this way.

5.2.2.2 Fixing layers : binders and tack coat material

Before dealing with the case of the interlayer products, it must be emphasized that the structural contribution of an overlay system is primary dependent on the way its different components interact.

If the adherence of the interlayer product is purely mechanical (nailed or by granular interlock), the effect is governed by granular friction and tensile resistance of the fixing materials. If the bond is made by a tack coat, binder or other viscoelastic medium, properties like binder viscosity and resistance to low temperature fracture are important. Experimental research has clearly shown to what extent the nature and quantity of the binder can influence the rate of crack propagation in overlays where nonwovens are applied [ref. 7].

5.2.2.3 Interlayer products

The variety of interlayer products available on the market is very wide and a large diversity can be found in their mechanical properties. Interlayer thicknesses of nonwovens and grids can range to about 2 mm. In comparison with overlay thicknesses of several cm, this is to be considered as an almost two-dimensional foil. Characteristic values of the product stiffness, stiffness modulus, ultimate strength and strain at ultimate strength are given in chapter 4. For some of these products, these properties are strongly anisotropic so that for two main directions different stiffness values have to be attributed to a same product. In the case of sand asphalt the characteristics needed are the same as for the asphaltic overlay (moduli, fatigue and crack propagation law).

We note that a given material will act as a reinforcement if its overall stiffness modulus is higher than that of the upper layer of the system. Owing to the fact that the bituminous overlay material is temperature susceptible, the ability of a given interlayer to reinforce will depend on the temperature. Therefore, a given product often acts as a reinforcement only in the medium and high temperature ranges, but not under very cold weather conditions.

Although much information is now available on the characteristics of the basic products of interlayer systems, it is noticeable that very little information is to be found on the characteristics of the interlayer once it is in place. This cannot be obtained by simply adding the properties of the individual components (for example tack coat + nonwoven or grid), but it is the kind of information which is required in order to get correct evaluations.

5.3 Performance laws

The development of a crack in a road structure may generally comprise of three steps involving different kinds of physical processes :

- initiation of the crack,
- propagation of the crack,
- further deterioration of the pavement after the crack has reached the surface.

At each step different physical laws can be applied according to the type of structure concerned and the predominant loading conditions applied to it.
For the onset and propagation of cracks in bituminous overlay materials the most important ones are the fatigue law and the crack propagation law.

5.3.1 THE FATIGUE LAW

This law allows the estimation of the number of loads N needed to initiate a crack resulting from the repetition of loads, either applied at a constant strain (ϵ) level or constant stress level :

$$N = \left(\frac{C}{\epsilon}\right)^m$$

The parameters C and m can be experimentally determined on the basis of repeated bending tests. For straight-run bitumen they can be readily estimated from parameters defining the volumetric mix composition and the binder rheological characteristics [ref. 3, 4].

5.3.2. THE CRACK PROPAGATION LAW

In case of a crack propagation analysis, the rate of crack growth in the overlay can be predicted using the empirical power law developed by Paris and Erdogan [ref 8], which relates the stress intensity factor (K) to the crack propagation speed (dc/dN) as follows :

$$dc/dN = A\,(K_{eq})^n$$

where A and n are fracture mechanics parameters.

The stress intensity factor K is depending on the geometry of the specimen (pavement), on the mode of opening of the crack (mode I or II) and on the crack length c.

For thermal movements of pavement materials only the opening mode I is most likely to occur. Traffic loads will introduce more complicated situations involving combinations of modes I and II (shear mode) [ref. 9].

The calculation procedure implies a series of successive analyses, in each one the stress intensity factors are computed. Hence an estimation can be made of the number of load repetitions N_f needed to propagate a crack through the overlay thickness h by integration [ref. 2, 8, 10]:

$$N_f = \int_0^h \frac{dc}{A \cdot (\Delta K(c))^n}$$

The normal way to experimentally determine the material parameters A and n is to examine the stable crack growth through asphalt beam specimens under repeated loading conditions [ref. 2, 11].

Different set-ups are possible, it is essential to measure the crack length c during the test over a sample geometry which is simple enough to allow an accurate estimation of the stress intensity factor K. Although this approach is an excellent way to describe the reflection crack problem, it remains rather unpopular for the reasons that the input data needed for its implementation are very scarce and difficult to obtain by experimental means.

Some equations derived by Shapery [ref. 12] can be used to determine the fracture mechanics parameters without performing expensive fracture tests, provided certain material properties are known.

Paris crack propagation law assumes that once a small crack is created, for instance through fatigue, it will propagate as a simple plane discontinuity through the material. However, there is no unanimity as to the validity of this theory in the case of heterogeneous materials such as bituminous mixes. Moreover, observations by Jacobs [ref. 10] have shown that a microcrack zone precedes the onset of macrocracks and that more fundamental principles such as the rate theory [ref. 13] should be applied.

5.4 Design Models

Table 5.2 gives a short review of the models used for the structural design of overlay systems.

Table 5.2a Overview of models for reflective cracking.

Model	Short description	References
A	Multilayer linear elastic	14, 15, 16, 17
B	Extended Multilayer	18, 19
C	Equilibrium equations	20
D	Mechanistic empirical overlay design method	2
E	Finite element/finite difference analysis + fracture mechanics	21 to 29
F	Blunt crack band theory	31

Table 5.2b Possibilities of the models.

	Model	A	B	C	D	E	F
Function	Stress/strain analysis	X	X			X	X
	Service Life prediction	X	X	X	X	X	
	Overlay design	X	X	X	X	(X)	
	Performance based comparison	X	X	X		(X)	(X)
Computing tool	Analytical	X	X	X	X		
	Finite Elements				X	X	(X)
	Finite differences					(X)	(X)
Dimensions	2D			X	X	X	X
	3D	X	X			(X)	
Loading conditions	Traffic	X	X	X	X	X	
	T.shrinkage base layer			X	X	X	X
	T.shrinkage upper layer			X	X		
	Warping				X	(X)	X
Damage law or criterium	Fatigue	X	X	X		(X)	
	Crack propagation		X		X	X	
	Thermal cracking			X			
Structure	Multilayer	X	X			X	
	Beam on foundation			X	X		
	With interlayer	(X)	(X)	X		(X)	X

(X) : at this moment, only in some cases.

5.4.1 MULTILAYER LINEAR ELASTIC MODEL (A in table 5.2)

The multilayer linear elastic model has been used for many years. The stress analysis used in these models is based on the generalized elastic layer theory of Burmister [ref. 14].This method works under the assumption that the structure is continuous through all of its layers and that it also displays the following properties :

- axi-symmetrical geometry,
- homogeneous, isotropic linear elastic materials,
- all layers extend to infinity in the horizontal plane,
- the friction between layers is either slippery or rough.

The parameters necessary to describe the structure are :

- number of layers,
- thickness of each layer,
- interlayer conditions between the successive layers,
- elastic properties of the individual layers (stiffness modulus and Poisson's ratio).

The fatigue phenomenon is currently used as the fundamental criterium in the design of new road structures for the calculation of the overlay thicknesses [ref. 14, 15, 16,17].

This model has been successfully used for many years as the basic element of many structural design methods and is now available under the form of software programs [ref. 17, 18]. However, discontinuities such as cracks, cannot be simulated with this model. It addresses exclusively the initiation phase of the cracking process in the case of an initially homogeneous and continuous pavement structure. The adjustment, for practical purposes, to the actual performance of roads was first made by adjustment factors. Before dealing with models including discontinuities it is worth remembering that the multilayer models are able to give in a first instance an evaluation of the original state of the structure to be treated (i.e. in the absence of any discontinuity in the lower layers).

5.4.2 APPLICATION OF THE LINEAR ELASTIC MULTILAYER THEORY TO CRACK PROPAGATION PROBLEMS (B in table 5.2)

Although linear elastic multilayer theories are not suitable to handle the case of localized damage features such as cracks, different trials have been made to include the crack propagation in a simplified manner.

The approach proposed e.g., in the MOEBIUS software [ref. 18] considers the pavement as initially sound. The structure is divided in as many layers as possible each having the initial properties of new asphalt. The first crack at the bottom is supposed to be initiated by fatigue. After the crack initiation stage the properties of the different sublayers are progressively reduced from the bottom to the top according to a rate of propagation determined from the knowledge of Paris'law. Such a procedure can obviously not claim to perfectly model the complexity of the cracking phenomena. The accuracy of the predictions are of course limited by the oversimplification and are also dependent on the input data.

Another trial to use a linear elastic multilayer program for the propagation of cracks from an old pavement through a new overlay was made by Van Gurp and Molenaar [ref. 19].

It is clear however that this type of extension of the elastic multilayer theory is merely a way to use an existing tool in a field for which it was not initially developed.

5.4.3 MODELS BASED ON EQUILIBRIUM EQUATIONS (C in table 5.2)

A procedure for the design of bituminous overlays on existing portland cement concrete (PCC) has been developed for the Arkansas State Highway transportation department [ref. 20]. The procedure is based on a simple mechanistic approach in which the main features are the movements of the concrete slabs close to joints or cracks and the thermal movements. Equilibrium equations were used for estimating the stresses, which were then used in a fatigue type of approach to estimate the lifetime of an overlay with or without a functional layer. This procedure has been implemented under the form of software and charts for practical overlay design.

5.4.4 MECHANISTIC EMPIRICAL OVERLAY DESIGN METHOD (D in table 5.2)

In another approach, Jayawickrama and Lytton [ref. 2] have developed a set of mechanistic empirical overlay design procedures to address reflective cracking in overlays of existing asphalt or portland cement concrete pavements. The basic design equations of this procedure are based on fracture mechanics and beam on elastic foundation concepts. They address fracture in the slab due to bending and shear caused by moving wheel loads and due to opening caused by thermal movements of the cracked pavements. The practical overlay design procedure was finally obtained by calibration of the mechanistic equations with performance data from in-service pavements.

This very powerful and practical tool allows the design of classical overlay solutions. It can now be applied for overlays on damaged asphalt pavements as well but, so far, it does not cover the case of interlayer products. Anyway, it can be used as a first step in the overlay design for the determination of the minimum reference thickness of a conventional type of overlay. The design software ODE (standing for Overlay Design Equations) which was developed for the implementation of this procedure, is restricted to six climatic regions of the United States. Its extension to other climatic regions requires the development of regression equations based on field data collections.

5.4.5 FINITE ELEMENT ANALYSIS (E in table 5.2)

The use of the finite element method for structural design calculations has become very popular during the last years. The success of this method is to a large extent related with the availability of computers powerful enough to solve the large systems of equations in a minimum of time. Finite element analyses are used for modelling of crack initiation as well as propagation.

For the initiation phase of a crack in the overlay, the lifetime of the overlay is determined via computation of the tensile strain at the bottom of the overlay and subsequently making use of a fourth power fatigue law. For the propagation phase of the crack through the overlay, analyses are performed by using fracture mechanics principles.

5.4.5.1 Some results for modelling of crack initiation

De Bondt [ref. 28] studied the effect of slab length and overlay thickness on the generated strain in the overlay above a crack/joint for a given temperature drop. He found that the product stiffness should be as high as possible in order to decrease the strain as much as possible.

At the Belgian Road Research Centre, 2D-computer simulations were carried out of overlaid cement concrete structures [ref. 21]. The effect of a given horizontal opening of the crack in the cement concrete base, which is the result of a temperature drop, is studied. A systematic analysis has been performed of the influence of the crack width, the overlay thickness and interlayer product on the induced horizontal strains in the overlay. For the effect of thermal loading, it was found that :

- the use of an interlayer product reduces the large strains occurring in the overlay close to the crack tip,
- the overlay thickness has only a limited influence on the resulting strains for a given horizontal opening of the crack,
- the best solution is obtained by the combination of a soft interlayer (allowing an easy horizontal movement) and a strengthening material at the base of the overlay.

Two- and three-dimensional finite element computer simulations were performed by Vanelstraete and Francken to simulate the effect of traffic on overlaid cement concrete slabs [ref. 21, 22, 23]. Symmetrical and asymmetrical loading positions with respect to the crack were considered.

Symmetrical loading position

A comparison was performed between 2D- and 3D-modelling, and it was concluded that 3D-modelling is necessary for a correct description of the problem [ref. 22]. The study led to the following conclusions :

- for regions very close to the crack tip (< 5 mm), the strains in the overlay are reduced by the presence of an interlayer system with comparable or higher stiffness than the asphaltic overlay. These interlayers lead to a considerable improvement in crack initiation time and are effective for the prevention of crack initiation, provided the interlayer system itself can withstand the strains induced in it,
- for distances further from the crack tip, interlayer systems do not reduce the horizontal strains and the crack initiation time considerably. The overlay thickness and overall design of the road structure is predominant here.

Asymmetrical loading position

Three-dimensional computer simulations of actual road structures were carried out in order to study the influence of different parameters on the vertical movements at the edges of overlaid cement concrete slabs (so-called "rocking effect") [ref.23] : type of base and subbase, stiffness and thickness of the layers, type of interlayer system.

From that study it is concluded that :

- rocking of concrete slabs is largely determined by the type and state of the soil and the bearing capacity of the base/subbase,
- interlayer systems can decrease slab rocking, mainly close to the crack tip. Their effect is however limited compared to that of the overlay thickness. This is illustrated in figure 5.4, in which the shear strain ϵ_{zx}, representative for the slab rocking, is given at various depths in the overlay for several overlay thicknesses and several interlayer systems,
- in case of severe slab rocking (e.g., if no base/subbase are present) and for the road structure and loading conditions described in ref. 23, an increase of the overlay thickness by 20 mm increases the lifetime before crack initiation by a factor of 2.2 ; a 2 mm interlayer system gives an increase of about 1.8.

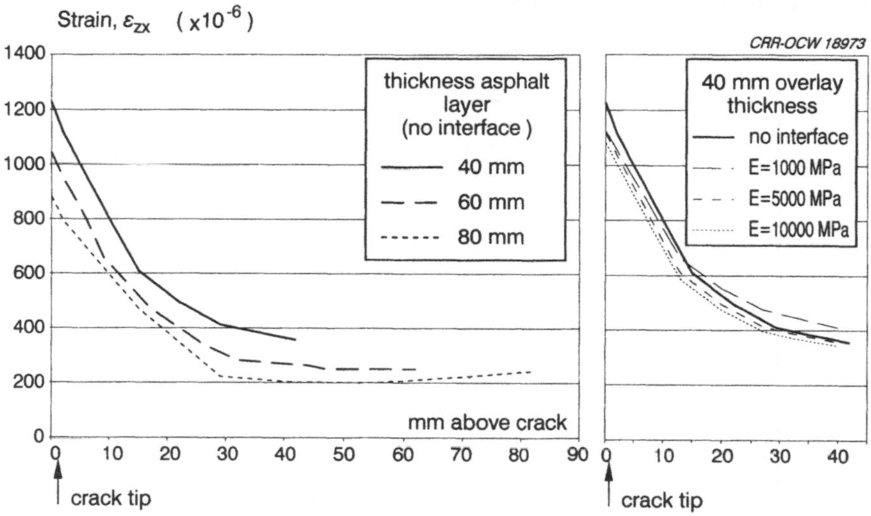

Fig. 5.4 Example of results for crack initiation as a result of traffic, obtained from finite element analysis [ref. 23]. Shear strain ϵ_{zx} (representative for slab rocking) at several depths in the overlay, for several overlay thicknesses and interlayer systems for the road structure and loading conditions described in [ref. 23].

5.4.5.2 Results of modelling of crack propagation

A large contribution to the introduction of the fracture mechanics concepts in road structures was performed by Majidzadeh [ref. 24, 25] in an attempt to explain the fatigue phenomenon.

The use of finite element calculations in conjunction with the fracture mechanics approach needs the use of special crack tip elements in order to get the correct values of the stress intensity factors of the Paris law. The first extensive trials in this field concern studies carried out by Monismith et al [ref. 26] for the case of soft interlayer systems.

The last years, PC-based finite element systems, such as CAPA [ref. 27], particularly suited for the modelling of crack propagation, were developed.

The overlay thickness is a predominant factor for the crack propagation. A larger overlay thickness not only implies that the crack has to grow over a longer distance before it reaches the overlay surface, but it also reduces the magnitude of the stress intensity factor at the tip of the crack. It must be reminded that the magnitude of this thickness effect depends on the site circumstances ; in case of poor soil support the effect of overlay thickness is of much more importance than in case of well-supported pavements. As an example, figure 5.5 shows for the situation of a transverse cracked flexible rural road in a soft soil area, the relative lifetime of a conventional overlay versus its thickness [ref. 28]. It is clear that a larger overlay thickness not only implies that the crack has to travel over a longer distance before it reaches the overlay surface, as indicated in figure 5.5 by the dotted line, but it also causes the reduction of the magnitude of the stress intensity factors at the tip of the crack; the difference between the upper curve and the dotted line. All in all, it can be seen that a thickness increase results in a non-proportional increase in lifetime.

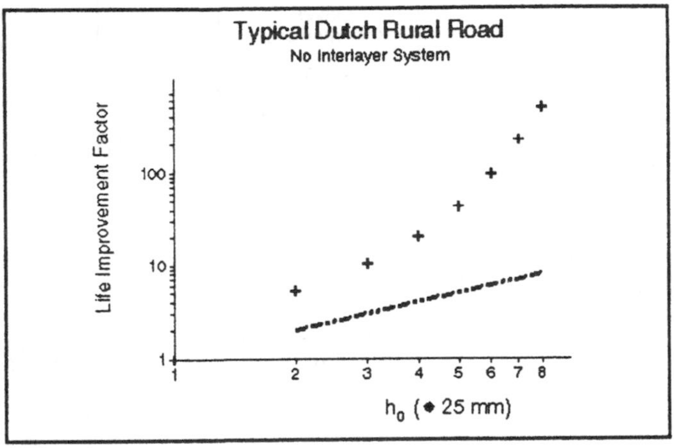

Fig. 5.5 Effect of thickness on overlay lifetime for specific conditions [ref. 28]. (Dotted line : effect of thickness only, without taking into account the reduction of the stress intensity factors with increase of thickness; upper curve : total effect of increase in thickness on the relative lifetime).

The modelling of the reflection of transverse shrinkage cracks made by Marchand and Goacolou for bituminous overlays on cement stabilised base layers under the effect of traffic and thermal stresses has shown that the path followed by a crack can be influenced in different ways by the geometry of the structure and the component properties [ref. 29]. This study also showed that the debonding mechanism leads to the development of horizontal cracks while strong bonds lead to a vertical propagation mechanism.

For the case of traffic loads de Bondt [ref. 28] found that the lifetime of an overlay which is placed on a cracked or jointed pavement can be improved by using reinforcement. It is clear that a proper reinforcing system enables the transfer of tensile forces at the bottom of the overlay after failure of the asphaltic mix has occurred at this location.

Also according to de Bondt [ref. 28], the reinforcement must be capable of generating a large force in order to be effective ; in other words, its pullout restraint must be high. The resistance to pullout not only depends on the stiffness of the reinforcement product, but also on the bond which is developed along the product. The product stiffness of interlayer products can be obtained from tensile testing (see 4.1.1.4). Bond stiffnesses can be determined by pullout tests (see 4.2.1.2). Figure 5.6 shows, as an example for a transverse crack in a typical semi-rigid pavement structure, the effect of the product stiffness S_{RF} and bond stiffness D_{tt} (RF-Bond) on the relative lifetime of the overlay. The lifetime for the unreinforced situation has been set to 1. It can be seen that the beneficial effect of a reinforcing system increases with increasing product stiffness as well as bond stiffness.

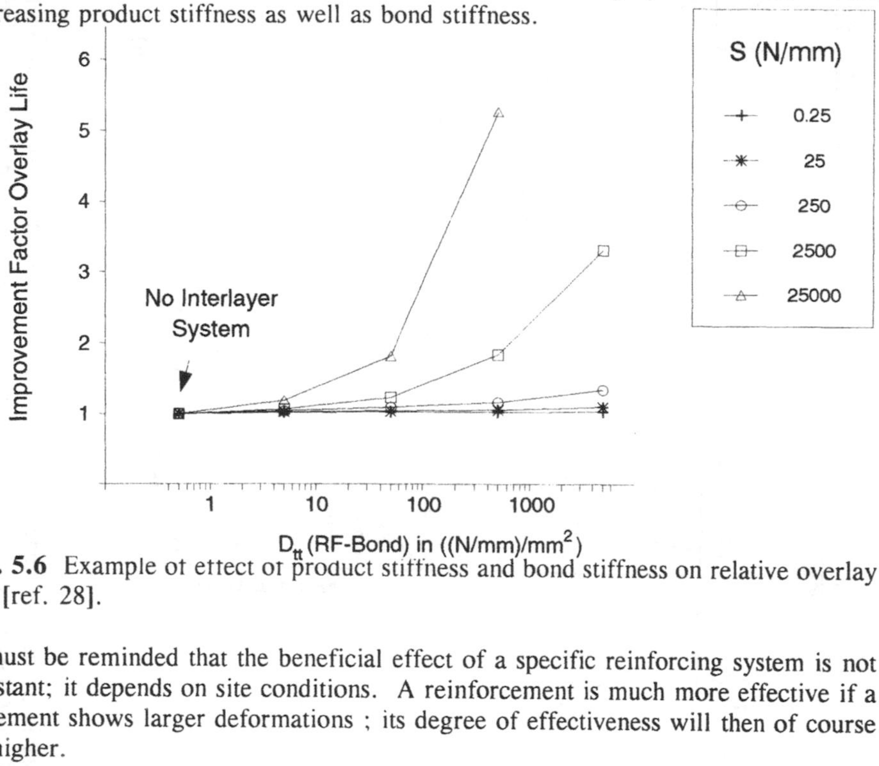

Fig. 5.6 Example of effect of product stiffness and bond stiffness on relative overlay life [ref. 28].

It must be reminded that the beneficial effect of a specific reinforcing system is not constant; it depends on site conditions. A reinforcement is much more effective if a pavement shows larger deformations ; its degree of effectiveness will then of course be higher.

Nomographs were deduced by Strauss et al in [ref.30] from finite element calculations, showing the influence of the overlay thickness, crack width and vertical (or horizontal) crack movement on the shear (or tensile) stress in an overlay under traffic (or thermal) loads. Figure 5.7 gives an example of such nomographs for the case of thermal loading (horizonal movement) under the assumption of perfect adhesion between underlayer and overlay (fig.5.7(a)) and when a soft interlayer system (such as SAMI or impregnated nonwoven) is applied (fig.5.7(b)). The influence of the horizontal crack movement, overlay stiffness, crack width and overlay thickness is shown.

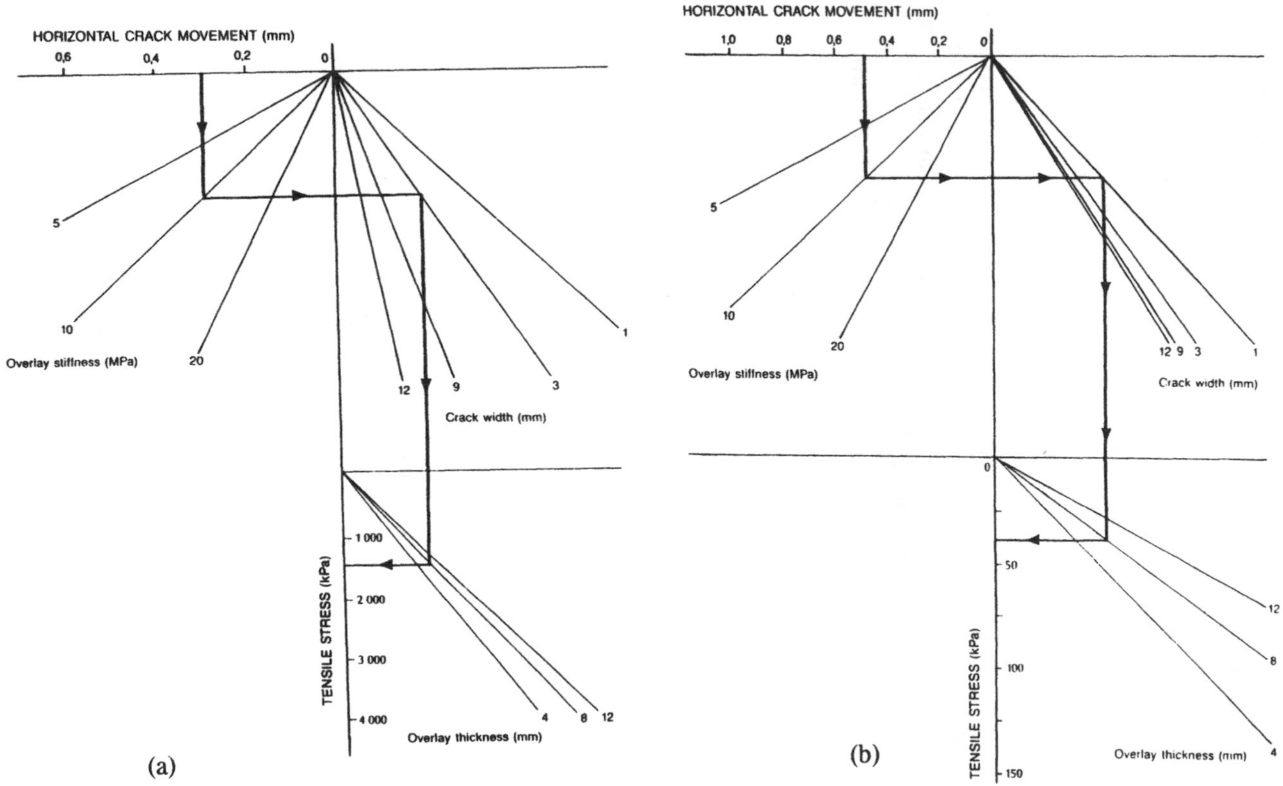

Fig. 5.7 Tensile stress in an overlay under horizontal movement (thermal load) : (a) for perfect adhesion between underlayer and overlay; (b) in case of a soft interlayer system (such as SAMI or impregnated nonwoven) [ref. 30].

From this figure, it can be concluded that :

- the use of a soft interlayer system leads to a significant decrease of the tensile stress in the overlay.
- the influence of material stiffness, horizontal crack movement and crack width on the stress in the overlay depends on the adhesion between overlay and underlying layers.
- the overlay thickness itself is not of great significance.

The authors however emphasize that the nomographs only serve to illustrate the relative importance of the different parameters. Using them in any circumstances, without being aware of the assumptions which were made, may result in large inaccuracies.

5.4.6 THE BLUNT CRACK BAND THEORY (F in table 5.2)

Another way to model the behaviour of cracked structures has been used by Haas [ref. 31]. Unlike theories based on the fracture mechanics approach these models simulate a crack as a broad zone of weak material. It can be considered indeed that an actual crack in an heterogeneous material such as a bituminous mix does not correspond with the theoretical representation which is assumed in the fracture mechanics approach. Moreover this crack morphology can be widely variable depending on the grading of the mix, the mechanism that generated the crack and the temperature at which fracture occurred.

To take account of the three dimensional nature of a damaged zone, the blunt crack theory represents an effective crack as a vertical band having a width close to the maximum grading size and to which (in a way similar to the extended elastic layered approach, model B of table 5.2), a weak modulus value is attributed. Such structures can be modelled either by finite elements or by finite difference computer programs [ref. 31]. The approach leads to a realistic evaluation of the stress distribution in the cross section of a cracked structure containing different types of interlayers.

The procedure is valuable in screening alternative treatments, but it is unable so far to describe the crack propagation mechanism.

5.5 Some remarks about modelling

The experience gained in the field of modelling has led to adopt finite element or finite difference programs as "the" stress-strain computation tools for the study of the reflective cracking phenomenon. However, it is clear that we have to deal with a very complex problem and we should be aware of the fact that we are still trying to simplify it to keep it viable for practical purposes. Hence some aspects of these methods still need care and attention. They are described below.

5.5.1 LIMITATIONS OF TWO-DIMENSIONAL MODELLING

As was already mentioned by Monismith [ref. 26], ideal solutions would consider a three dimensional pavement system. It appears that in the case of traffic loading the plane strain assumption generally implicitly adopted in 2D-modelling does not correspond to reality and may even lead to wrong statements [ref. 21].

5.5.2 THE BOUNDARY CONDITIONS

Finite element programs model structures with limited spacial dimensions, they are therefore not directly usable for continuous pavement layers, unless boundary conditions are introduced to simulate the reaction forces of the remaining adjacent parts of the structure. Away from the crack, these conditions can be evaluated more precisely by a preliminary study using a linear elastic model of the same structure without any crack.

5.5.3 UNCONVENTIONAL MECHANICAL PROPERTIES

Much needs to be done in order to improve the laws describing the behaviour of functional interlayers because they are, even more than the other components, non-linear, non-elastic and non-isotropic.

5.6 Conclusions

Finite element modelling has become a powerful tool for the design and evaluation of overlay systems. Models and software tools specifically developed to study the phenomenon of reflective cracking exist and are still being improved. These models allow a better insight in the functioning of overlay systems and their components.

Although a lot of interesting work has been done, large efforts are needed in this field to come to real practical solutions for the design of overlay systems. Only tentative results are given so far, describing particular situations under specific loading conditions. It is not clear whether the results obtained by different authors are comparable and can be extrapolated to other cases. The situations being studied so far either concern the crack initiation or the crack propagation phase, but do not describe the whole cracking phenomenon. Comparisons are necessary to compare the calculated relative lifetimes with those in the field.

5.7 References

1. G. Colombier : "Fissuration des chaussées, nature et origine des fissures, moyens pour maitriser leur remontée". Proceedings of the 1st RILEM-Conference on Reflective cracking in pavements. Liège, pp. 3-22, 1989.

2. P.W. Jayawickrama, R.L. Lytton : "Methodology for predicting asphalt concrete overlay life against reflection cracking". Proceeding of the 6th International Conference on Structural Design of Asphalt Pavements. Ann Arbor, pp. 912-924, 1987.

3. F. Bonnaure, G. Gest. A. Gravois and P. Ugé : "A new method of predicting the stiffness modulus of asphalt paving mixtures". Proceedings A.A.P.T., 1977.

4. L. Francken and C. Clauwaert : "Characterization and structural assessment of bound materials for flexible road structures". Proceedings of the VIth Int.Conf on the Structural Design of Asphalt Pavements. Ann Arbor, Michigan, pp. 130-144, 1987.

5. L. Francken and A. Vanelstraete : "Complex moduli of bituminous binders and mixes : Interpretation and evaluation". Eurasphalt & Eurobitume Conference, Paper 4.047, Strasbourg, 1996.

6. W. Arand : "Behaviour of asphalt aggregate mixes at low temperatures". Proceedings of the fourth RILEM symposium on Mechanical tests for bituminous mixes. Budapest, pp. 68-84, 1990.

7. J. M.Rigo et al.: "Laboratory testing and design methods for reflective cracking interlayers". RILEM Conference on Reflective cracking in pavements. Liège, pp. 79-87, 1989.

8. P.C. Paris and F. Erdogan : "A critical analysis of crack propagation laws". Journal of basic engineering. Transaction of the american society of mechanical engineering. Series D, vol.85, pp. 528-553, 1963.

9. J. Rosier, C. Petit, E. Ahmiedi, A. Millien : "Mixed mode fatigue crack propagation in pavement structures under traffic load". Proceeding of the 3rd International RILEM-Conference on Reflective Cracking in Pavements, Maastricht, pp. 143-152, 1996.

10. M. Jacobs : "Determination of crack growth parameters of asphalt concrete, based on uniaxial dynamic tensile tests". Proceedings of the fourth RILEM symposium on Mechanical tests for bituminous mixes. Budapest, pp. 483-496, 1990.

11. A.A.A. Molenaar : "Structural performance and design of flexible road constructions and asphalt concrete overlays". Ph. D-Thesis T.U. Delft, 1983.

12. R.A. Schapery : "A method for predicting crack growth in non homogeneous viscoelastic media". Int. Journal of fracture mechanics, pp. 293-309, 1978.

13. A. Tobolski and H. Eyring : "Mechanical properties of polymeric materials" Journal of Chemical Physics 11, pp.125-134, 1943.

14. J. Verstraeten : "Stresses and displacements in elastic layered systems" Proceedings of the 2nd Int.Conf on the Structural Design of Asphalt Pavements. Ann Arbor, August, 1967.

15. A.I. M.Claessen, J.M. Edwards, P. Sommer and P. Ugé : "Asphalt Pavement Design". The Shell method. Proceedings of the IVth Int.Conf on the Structural Design of Asphalt Pavements. Delft, pp. 39-74, July 1977.

16. B. de la Taille, P. Schneck and F. Boudeweel : "ESSO Overlay design system". Proceedings of the Vth Int.Conf on the Structural Design of Asphalt Pavements. Delft, pp. 682-694, July 1982.

17. Valkering and D.R. Stapel : "The Shell pavement design method on a personal computer". Proceedings of the 7th International Conference on Asphalt Pavements. Nottingham, Vol 1 pp. 351-374, 1992.

18. B. Eckman : "ESSO MOEBIUS Computer software for pavement design calculations". User's manual. Centre de Recherche ESSO. Mont Saint Aignan, France, June 1990.

19. C.A.P.M. Van Gurp and A.A.A. Molenaar : "Simplified method to predict reflective cracking in asphalt overlays". RILEM Conference on Reflective cracking in pavements. Liège, pp. 190-198 ,1989.

20. S.B. Seeds, B.F.Mc Cullough and F. Carmichael : "Asphalt concrete overlay design procedure for portland cement concrete pavements". Transportation Research Record 1007, Washington D.C., pp. 26-36, 1985.

21. L. Francken and A. Vanelstraete : "Interface systems to prevent reflective cracking". Modelling and experimental testing methods. 7th ISAP conference on asphalt pavement. Nottingham, vol1 pp. 45-60, 1992.

22. A. Vanelstraete and L. Francken : "Numerical modelling of crack initiation under thermal stresses and traffic loads". Proceedings of the 2nd International RILEM-Conference on Reflective cracking in pavements, pp. 136-145, 1993.

23. A. Vanelstraete and L. Francken : "Laboratory testing and numerical modelling of overlay systems on cement concrete slabs", Proceedings of the 3rd RILEM-Conference on Reflective Cracking in Pavements, pp. 211-220, 1996.

24. K. Majidzadeh, E.M. Kaufmann and D.V. Ramsamooj : "Application of fracture mechanics in the analysis of pavement fatigue". Proceedings A.A.P.T. vol 40, pp. 227-246, 1970.

25. K. Majidzadeh, L.O. Talbert and M. Karakouzian : "Development and field verification of a mechanistic structural design system in Ohio". Proceedings of the IVth Int. Conf. on the Structural Design of Asphalt Pavements. Delft, pp. 402-408, July 1977.

26. C.L. Monismith : "Reflection cracking. Analyses, laboratory studies, and design considerations". Proceedings A.A.P.T. vol 49, pp. 268-313, 1980.

27. A. Scarpas, J. Blaauwendraad, A. H. de Bondt, A.A.A Molenaar, CAPA : "A Modern tool for the analysis and design of pavements". Proceedings of the 2nd International Conference on Reflective Cracking in Pavements, pp. 121-128, 1993.

28. A.H. de Bondt : "Anti-Reflective Cracking Design of (Reinforced) Asphaltic Overlays". Ph. D.-Thesis, Delft University of Technology, 1997.

29. J.P. Marchand and H. Goacolou : "Cracking in wearing courses". Proceedings of the Vth Int.Conf on the Structural Design of Asphalt Pavements. Delft, pp. 741-757, July 1982.

30. P. J. Strauss, E. Kleyn, J.A. du Plessis : "Field performance, laboratory testing and predictive models for modified binders used in reflection cracking". Proceedings of the 7th International Conference on Asphalt Pavements, Vol. 3, pp. 341-355, 1992.

31. R. Haas and P.E Joseph : "Design oriented evaluation of alternatives for reflection cracking through pavement overlays". RILEM Conference on Reflective cracking in pavements. Liège, pp. 23-46, 1989.

6

Implementation of overlay systems on the construction site

F. Verhee, J. P. Serfass, T. Levy

6.1 Application of interlayer systems and wearing courses

6.1.1 PREPARATORY WORKS

These are intended to make adhesion to the underlying layer possible and to homogenize this adhesion.

6.1.1.1 Cleaning the underlying layer
The existing road surface must not only be cleared from dirt, but also be made sufficiently rough to ensure adhesion. Any bleeding of binder or any thermoplastic road marking protruding from the surface requires a particular treatment (e.g., burning, scraping, chipping, ...).

6.1.1.2 Regulating
The laying and/or good performance of the interlayer system and the wearing course often require a uniform thickness (e.g., for sand asphalt, too small a thickness reduces its effectiveness and an excessive thickness under a thin wearing course may cause rutting) and the absence of discontinuities (e.g., a discontinuity in the underlying surfacing may prevent the adhesion of an impregnated nonwoven).

Acceptable deformations, in particular sudden differences in level such as slab stepping, vary according to the technique : from zero for grids to several centimetres for thick wearing courses. It is important that no voids are present under the interlayer system. Otherwise, these may remain after overlaying or can lead to unadequately compacted spots.

Usable techniques are either regulating (adding materials) or milling (removing materials).

Protruding road markings can also be removed by burning or shotblasting.

6.1.1.3 Treating cracks
Cracks wider than 5 mm should be sealed to waterproof them and hold their edges together. This is done with a bituminous mastic, after the crack has been cleaned and dried (see 6.1.2). The seal need not be chipped if it will not be trafficked.

Prevention of Reflective Cracking in Pavements. Edited by A. Vanelstraete and L. Francken. RILEM Report 18.
Published in 1997 by E & FN Spon, 2–6 Boundary Row, London SE1 8HN. ISBN 0 419 22950 7.

Generally, when the crack edges have started to spall off, a local repair is necessary. This consists of restoring the wearing course, or even the base course, over a width of 50 cm on either side of the crack, after cutting out and removing the existing bituminous material. Of course, the pavement to be overlaid must be free from weak spots in the road base and subbase.

6.1.1.4 Cement concrete pavements [ref. 1]

On cement concrete pavements subjected to severe vertical movements at joints ("slab rocking"), it may be desirable to reduce those movements in order to limit the thickness of the bituminous overlay. Several techniques are available :

a. If slab rocking is very severe (relative movements of 0.5 mm and more), the slabs may be cracked in place into 2 to 6 pieces using a power hammer or a "guillotine", after which the pieces of slab are "seated" into the subbase by means of a heavy roller. Thus stabilized, the pieces of slab will present only minor vertical and horizontal movements. Regulating will generally be necessary. The cracked slabs can be used as a road base. The bituminous overlay will still need to be thick. On minor jobs, the slabs may be sawn through, in which case they will most often be cut into only two parts.

b. When slab rocking is of the order of 0.25 to 0.50 mm, the relative vertical movements can be mitigated by improving load transfer across the joints or by reducing the cavities under the slabs.

   - Improving load transfer : this is achieved by installing dowels or connectors at transverse joints.

     - Dowels are installed in longitudinal slots, which are generally made by sawing. They are fixed with a mortar the binder of which generally contains resins. The dowels should permit horizontal movements to allow for expansion or contraction.
     - Connectors are cylinders formed by two semicircular shells. One shell is glued to the approach slab and the other to the leave slab, with resin. They are keyed together so as to prevent only vertical movements, leaving the shells independent in the horizontal direction.

       Installation is by drilling cylindrical holes to both sides of the joint and inserting the connectors after gluing. Generally, four connectors are used per lane.

   - Improving the subbase by filling the cavities resulting from mud pumping is done by injecting a cementitious slurry of cementitious binder under the slab. The slurry is generally injected through five cylindrical holes, one in each corner and one at the centre of the slab. This technique requires delicate workmanship to reduce poor bearing conditions (considering slab camber, limiting the injection pressure to avoid breakage).

6.1.2 SEALING UNDER BITUMINOUS WEARING COURSES OR SURFACE DRESSINGS

6.1.2.1 Objectives

Crack sealing (figure 6.1) before laying the wearing course has three objectives :

- to waterproof the crack,
- to hold the edges of the crack together by filling up the gap between them,
- to improve the resistance to cracking of the wearing course by adding an extra amount of modified bituminous binder at the bottom of this course where it will be most severely stressed. The sealing compound will, indeed, become fluid and migrate to the bottom of the bituminous overlay while the latter is being laid. We note however that some sealants for crack sealing react with the overlay particularly if the overlay contained modified binder.

Fig. 6.1 Sealing under a bituminous wearing course or a surface dressing waterproofs the underlying layer and makes it richer in binder, thus delaying reflective cracking.

6.1.2.2 Technique

There are two techniques :

- one, called filling, consists of routing out the crack for example by sawing or milling and then filling it with bituminous mastic (modified bitumen, sometimes slightly admixed) while levelling the seal off flush with the road surface. It is seldom used, as it is expensive and difficult to implement ; moreover, it adds little binder to the bottom part of the asphalt course,
- the other, called bridging, consists, after preparing the surface, of spreading in and on either side of the crack a mastic to a height slightly overtopping the existing surface, and sanding the seal if necessary to prevent it from sticking to vehicle tyres. This technique is very widespread and is discussed further below.

6.1.2.3 Sealing by bridging

- Preparing the crack

  The crack must be clean and dry to allow good adhesion of the mastic. This is generally achieved by using a hot-air blower.

- Applying the sealing compound

  The sealant is most often a bitumen highly modified with polymers. The compound should have been tested for sticking power and elasticity. A primer is often necessary to promote adhesion. The mastic is applied by means of a casting shoe, to a height 1 to 2 mm above the surface and over a total width of 5 to 15 cm centred on the crack. Microchippings are applied if there is a risk of the sealant sticking to the tyres of construction or normal traffic vehicles. This treatment also reduces the wearing away of the surface film by traffic.

- Wearing course

  Any type of wearing course can be used. However, the thinner the course, the higher the risk of the seal being reflected at the surface (with 4 cm or thicker courses of asphalt there is no more risk of reflection). This is generally only an aesthetic defect. Under a wearing course of constant thickness, it can be limited or prevented by varying the width of the crack-bridging seal or its height above the surface of the base course.

6.1.3 SAND ASPHALT UNDER BITUMINOUS WEARING COURSES

Sand asphalt (or bituminous mortars) (figure 6.2) are used for interlayer systems. They are always covered with at least one layer of bituminous concrete.

Fig. 6.2 The thickness of a sand asphalt interlayer systems - about 2 cm - accommodates surface irregularities. The material is laid as a conventional bituminous overlay.

A tack coat is required. The amount of tack coat (the "rate of application") generally lies between 300 and 500 g/m² of emulsion, depending on the condition of the underlying surfacing.

Modified bitumen-based sand asphalt can be manufactured in two ways :

- the modified binder is ready for use. It only requires an adjustment of temperatures for adequate viscosity at the time of coating ;
- the additive is mixed in the coating plant. This applies to fibres, as well as to polymers in certain techniques. The additive - in fusible bags or in bulk - is introduced before or during the injection of the bitumen. Various techniques involving timing, introduction and proportioning devices have been designed by suppliers,
- to some contractors, the very high binder content and the resulting workability problems are reasons to restrict compaction to smooth-wheeled rollers, only the joints being slightly vibrated. Others use tandem rollers only, but compact by vibration to cause binder bleeding at the surface.

Finally, some - fewer in number - use a combination of rubber-tyred and smooth-wheeled rollers. The difficulty here is to prevent sand asphalt from sticking to the tyres, which must be constantly sprinkled with a dispersion of oil and protected from cooling (burlaps absolutely necessary).

It is not advisable to store "anticracking" sand asphalt mixes in hoppers, where they may tend to agglomerate as a result of their high binder contents.

In each process the recommended temperatures should be selected, in view of the thinness of the layer and the workability of the material.

The following should be noted with respect to the laying of "anticracking" sand asphalt mixes :

- sand asphalt should be laid with the spreading screws of the finisher locked in the appropriate position ; when working with two finishers abreast, the first should operate with locked screws and the second with a feeler,
- laying speed may be very high (up to 15 m/min or even more),
- compaction methods vary from one contractor to another, depending on procedures,
- the joints are fairly easy to make and should be well sealed.

6.1.4 NONWOVENS UNDER BITUMINOUS WEARING COURSES OR SURFACE DRESSINGS

6.1.4.1 Weather conditions

Good weather conditions are required for installation. The minimum temperature is 5°C and the surface to be overlaid should be dry. Work should be interrupted in case of rain.

6.1.4.2 Tack coat (figure 6.3)

Use is made of bitumen emulsions (> 65 %) or hot-sprayed binders. The latter are generally recommended, for the following reasons :

- no flowing out (on sloping or milled road surfaces),
- the nonwoven can be laid immediately after the binder has been sprayed, whereas with emulsions it is generally recommended to allow some time for breaking before applying the nonwoven,
- they make a stronger bond to the underlying surface.

Fig 6.3 Impregnated nonwoven : the binder is most often applied hot as this virtually eliminates the risk of sticking to the tyres of site vehicles and machines traversing the nonwoven.

The appropriate grade of binder also depends on weather conditions. In cold weather, high-penetration and low-viscosity bitumens (e.g., pen. 180/200) are used to achieve good adhesion of the nonwoven ; on the other hand, harder binders are employed in hot climate regions to avoid impregnation of the nonwoven during laying (problems with material sticking to lorry tyres during overlaying). The effectiveness of the system can be improved by using polymer-modified bitumens.

The role of the bituminous binder is to bond the overlay to the underlying surface and to impregnate the nonwoven. Depending on the type of product and the porosity of the underlying surfacing, the total amount to be spread generally varies between 700 and 1200 g/m² of residual binder. With certain nonwovens the rate of spread may be as high as 1400 g/m². A special laying procedure is required in such cases : spraying the binder in two passes under and on top of the nonwoven + chippings to prevent sticking to the wheels of construction equipment. With emulsions the rate of spread is proportioned to leave the required amount of residual binder.

6.1.4.3 Nonwovens

The nonwovens used are most often needle-punched (mechanically bound), thermally bound or even woven polypropylene (sometimes polyester). The maximum width is usually limited to 4 m and the maximum weight of a single roll should not exceed 100 kg, to permit easy handling on site. The width, of course, depends on the dimensions of the carriageway. Rolls are stored in their original packings in a dry place, to avoid the absorption of moisture or mechanical damage.

Fig. 6.4 The application of a nonwoven requires special care to prevent wrinkling and creasing.

Nonwovens are laid (figure 6.4) preferably with an unrolling machine, or manually over small areas. To prevent wrinkling, they are unrolled with tension and pushed down by brushing from the centre to the sides. Any wrinkles, creases or folds are cut. Adjacent rolls placed side by side on the road are laid with a longitudinal overlap of 10 to 15 cm, and consecutive rolls with a transverse overlap of 10 to 20 cm ; with the beginning of each roll underneath the previous one, in the direction of paving for the overlay, to prevent "rucking up" in front of the finisher. Bituminous binder is applied at overlaps at a rate of 300 g/m².

Laying nonwovens in bends must be done carefully to avoid wrinkling. The products are cut over the full width and placed while following the curve of the bend as closely as possible. Overlaps are made as described above. Chipping is mandatory to prevent sticking.

Fig. 6.5 Abrupt manoeuvres in applying a bituminous wearing course on a nonwoven may lead to wrinkling and creasing.

6.1.4.4 Bituminous wearing courses (figure 6.5)
Road traffic must not be allowed on the installed nonwovens, to avoid any damage (e.g., "rucking up"). If traffic cannot be diverted, a bituminous mix or coarse aggregate must be spread for protection.

When using an emulsion, the bituminous wearing course must not be laid before the breaking of the emulsion is complete.

Abrupt or needless manoeuvres (braking, turning) should be avoided. A small quantity of bituminous material or coarse aggregate should be spread when a nonwoven impregnated with a bituminous binder sticks to the tyres of site lorries or finishers.

The heat from the bituminous mix softens the binder, thus contributing to the full impregnation of the nonwoven. The thickness of the wearing course depends on the condition of the existing pavement and the loads it has to carry ; a minimum thickness of 30-40 mm is recommended for standard mixes.

6.1.4.5 Surface dressings

Apart from precautions to limit wrinkling in the nonwoven during surface dressing, the latter is applied in the conventional way as shown in figure 6.6.

Fig. 6.6 Surface dressings on nonwovens are applied as conventional surface dressings.

6.1.5 MEMBRANE WITH THREADS SPRAYED IN PLACE UNDER BITUMINOUS WEARING COURSES AND SURFACE DRESSINGS

There are two types of techniques.

In the first, the membrane is manufactured on site and laid in a single pass, using a special machine (figure 6.7) which sprays an elastomer-bitumen emulsion and applies polyester threads on it in an interwoven pattern. The membrane is invariably spread with chippings for protection and then overlaid with a wearing course of surface dressing or bituminous concrete or a bituminous structural layer (in case of a new construction or strengthening). The chipping spreader used may be vehicle-mounted or self-propelled.

Fig. 6.7 Membrane with threads sprayed in place : the threads, wound on flexible reels, are ejected through nozzles onto the previously sprayed binder.

The rates of application are adapted to the condition of the existing surface and may range from :

- 1.5 to 1.9 kg/m² (residual binder : 1.0 to 1.3 kg/m²) for the emulsion,
- 80 to 120 g/m² for the threads.

The rate of chipping of the membrane depends on the type of wearing course as shown in table 6.1.

Table 6.1 Rate of chipping for membrane with threads sprayed in place.

Wearing course	Chippings (size and quantity)
Bituminous concrete	6/10 5-6 l/m²
Double surface dressing (aggregate sizes 10/14 + 4/6) (*)	10/14 9-11 l/m²
Double surface dressing (aggregate sizes 6/10 + 2/4) (*)	6/10 7-8 l/m²

(*) Measures in mm.

The second technique uses glass fibres. These are cut from of glass threads by a mechanism mounted on the machine that is used to spread them :

- either a binder sprayer : in this case, the fibres are dropped directly on the binder just sprayed on the existing surface, and chippings are spread shortly after,
- or a chipping sprayer : in this case, the fibres are injected in the flow of aggregates falling on the sprayed binder.

Like in the first technique, the structures and rates of application of surface dressings are adapted to the condition of the underlying surface.

Unlike manufactured nonwovens, this type of membrane can be laid directly on a deformed surface and raises no problems of wrinkling and creasing, even on bendy roads. Joint-making is no particular problem, as machine passes may overlap.

6.1.6 GRIDS UNDER BITUMINOUS WEARING COURSES

6.1.6.1 Preparatory works

These are similar as for nonwovens (see 6.1.1).

The substrate should be as even as possible ; a regulating course should be applied if necessary.

Before the grid has been laid in place, a layer of emulsion should be sprayed onto the surface to be treated. The minimum amount of residual bituminous binder is 400 g/m^2.

Self-adhesive grids are sometimes used. However, if the substrate is moist, a layer of emulsion will be necessary to avoid adhesion problems.

6.1.6.2 Laying the grid

Grids are almost invariably sandwiched between two layers of bituminous concrete.

There have been applications on cement concrete pavements, as far as their surfaces are free from serious irregularities ; if not, prior regulating is required.

Grids should be installed according to the manufacturer's instructions. The grid is unrolled and laid flat, and fixed mechanically at the beginning of each roll and between rolls. Overlaps are generally around 200 mm longitudinally and 100 mm transversely. Wide bends are taken by stretching out the grid : if this is not possible, the grid should be cut transversely and the edges fixed mechanically (see figure 6.8). To prevent wrinkling or creasing during the laying of the bituminous wearing course, the grid is stretched out and mechanically fixed.

Fig. 6.8 Elimination of folds in grids, e.g., in bends : cutting, superposing and fixing (OCW-CRR 3320/10A).

To prevent the grid from sliding under the finisher, successive rolls in the direction of laying should be placed with the beginning of each roll underneath the end of the previous one.

No traffic must be allowed on grids before the bituminous overlay is in place, except for site traffic laying the overlay. These vehicles should avoid braking, accelerating, or turning abruptly on the grid.

In some countries specifications require a protective layer of chippings to reduce the risks of displacement or wrinkling under site traffic.

6.1.6.3 Laying the bituminous overlay

The bituminous wearing course is laid as usually, with a minimum thickness of 40-50 mm, depending on the type of overlay. Any abrupt manoeuvre (braking, turning) during the installation on the grid may cause wrinkling or creasing and should, therefore, be avoided. If necessary, some bituminous material should be spread manually on the grid in front of the tyres of lorries and finishers, to prevent sticking.

The wearing course is compacted in the conventional way.

6.1.7 STEEL REINFORCING NETTINGS

6.1.7.1 Laying the netting

The steel reinforcing netting is unrolled, flattened with a rubber-tyred roller, and fixed to the underlying surfacing. This fixing generally takes the form of a combination of nailing at the beginning and end of each roll and full-width sandwiching in a bituminous slurry. Nailing is carried out with stirrup hooks and with nails of appropriate types and sizes. The maximum spacing between nails at the beginning and end of each roll is 0.5 m ; additional nails should be used at places where the grid is not in contact with the underlying surface.

In bends, the steel reinforcing netting should be cut and the edges laid upon each other, to enable the netting to follow the curvature of the road while remaining flat. The net should be fixed to the substrate at such places and any excess parts should be removed.

Connections between rolls are made by butting the nettings and fastening the ends. Overlaps in the longitudinal direction of the road should be at least 0.30 m.

The net should be cut along and around manholes, inspection chambers, gully gratings, air holes and other elements. A minimum distance of 5 cm should be observed from the edges of the obstacle.

6.1.7.2 Laying the bituminous slurry

Some types of steel reinforcing nettings are embedded in a bituminous slurry (see figure 6.9) ; others are installed by nailing. A tack coat (minimum amount of residual binder : 400 g/m^2) is applied before laying the slurry.

An elastomer-bitumen slurry with aggregate size 0/7 mm is spread at a minimum rate of 12 kg/m^2.

If the overlay is to be a single course of porous asphalt, two passes of slurrying are made, using a type 0/7 for the lower and a type 0/4 for the upper part.

Fig. 6.9 Laying a bituminous slurry on a steel reinforcing netting (OCW-CRR 3290/18A).

6.1.7.3 Laying a bituminous surfacing

A tack coat must be applied. The minimum thickness of standard bituminous overlay is 40-50 mm, depending on the type of overlay.

The first layer must not be compacted with a vibrating roller.

6.1.8 THREE-DIMENSIONAL HONEYCOMB GRIDS

6.1.8.1 Preparing the underlying surface

The surface is carefully levelled and regulated. Rutted or severely cracked areas are milled out.

Fig. 6.10 Bituminous material is applied in a single pass to a total thickness of 5 cm on a three-dimensional honeycomb grid.

6.1.8.2 Laying the grid

The elements of netting are arranged manually and linked together with keys (figure 6.10). Any extra widths are cut off in place. Where necessary, the reinforcing is bonded to the underlying surface by means of special staples which are fixed by nailing. Shorter elements are used in sharp bends. As a rule, the tack coat is applied by means of a binder sprayer, which can be allowed on the reinforcement without problems. Small areas or connections are finished off with a lance.

6.1.8.3 Laying the bituminous overlay

The bituminous overlay is applied by a finisher, which equally runs on the reinforcement (coated with emulsion). It is then compacted.

The bituminous overlay cover of the reinforcement should be at least 15 to 25 mm. With honeycomb cells of 30 mm in thickness, this means that the total thickness of the bituminous overlay should be at least between 45 and 55 mm.

6.1.9 THICK DRESSINGS (SAMI's) UNDER BITUMINOUS WEARING COURSES

A thick dressing, also known as SAMI (stress-absorbing membrane interlayer), is a layer of highly modified bituminous binder sprayed hot at a rate of 2 to 3 kg/m². The binder is spread with chippings to allow trafficking. Its specific features as compared to a conventional surface dressing are :

- a higher viscosity of the binder, which must be applied with specific or adapted equipment. Working conditions to ensure good adhesion are more critical, especially with respect to temperature and the absence of moisture from the substrate,
- a protective layer, as traffic would damage the layer of binder if it were allowed directly on it. Generally chippings are spread for use by site traffic. However, this involves a risk of the chippings fixing the membrane onto the underlying surfacing by "indentation", which may prevent the membrane from distributing longitudinal deformations at cracks as intended. To solve this problem, one firm has suggested to replace the chippings with a cold-laid bituminous slurry.

6.1.10 COMBINED PRODUCTS : GRID ON NONWOVEN

The interlayer system consisting of a grid on a nonwoven, should be installed according to the manufacturers instructions. Usually, the technique described in 6.1.4 for the nonwoven is applied.

6.1.11 OTHER INTERLAYER TECHNIQUES

Other interlayer techniques are also used, albeit on a smaller scale :

a. *local interlayer systems at cracks*, for example an impregnated nonwoven, a prefabricated polymer-bitumen strip, or a grid. The results have generally been unsatisfactory, since in addition to the crack two new discontinuities can be created that may induce further cracking ;

b. *full-width interlayer systems bringing in an additional amount of bituminous binder*, such as a slurry seal. But the amount of binder added is small (binder contents comparable to those of the underlying bituminous materials).

6.2 Wearing courses alone

Only the specific features of wearing courses applied on cracked road pavements are discussed here.

6.2.1 SURFACE DRESSINGS

If the underlying surface has been prepared correctly, the only adaptation which may be necessary is spraying the binder at a slightly higher rate in order to waterproof the cracks as well as possible. However, this higher rate must by no means entail a risk of bleeding of binder.

6.2.2 THICK DRESSINGS (figure 6.11)

These dressings, also known as SAMI's (stress-absorbing membranes) and similar to those defined in section 6.1.9, have been specially designed to control cracking by :

- using a highly elastic binder (highly polymer-modified bitumen),
- laying a thick course (2 to 3 kg/m^2).

Fig. 6.11 The binder of a thick dressing (SAM) is to be sprayed at a rate of 2 to 3 kg/m^2, on a perfectly clean surface.

The specific features have been described in section 6.1.9. Requirements for application (such as temperatures and no moisture on the surface to be overlaid ; the temperature of the binder, and also cleanliness of aggregates) are even more stringent, as the dressing is used as a wearing course in this approach.

Similarly, the chippings may have the same adverse effect as described above on the anticracking function of the dressing ("indentation").

6.2.3 CONVENTIONAL BITUMINOUS WEARING COURSES

First of all, it should be noted that any bituminous layer added to a road pavement will strengthen its structural bearing capacity and, hence, reduce the appearance of cracks. Furthermore, an increase in thickness of the wearing course will increase the distance and, consequently, extend the time for cracks to propagate to the surface.

By increasing the content of bituminous binder, using polymer-modified binders and reducing the size of aggregates, more flexible bituminous materials can be obtained with a higher resistance to cracking. However, these materials will be more susceptible to deformation (rutting). Adding some types of fibres also seems to have a beneficial effect on resistance to cracking.

6.2.4 SPECIFIC BITUMINOUS WEARING COURSES

The special measures to be taken in manufacturing and laying depend on the modification technology used. Figure 6.12 shows a particular view of equipment in use. For bituminous materials containing a ready-to-use modified binder, the special measures mainly relate to the storage and heating of the binder. Modified binders should be stored at a temperature which cannot affect their properties in any way (required storage temperatures may vary with storage time). Certain modified binders (rubber-bitumen, some polymer-bitumens) need to be kept under constant agitation, to prevent any segregation of phases. When coating aggregate, the binder must have a viscosity enabling it to be properly pumped and injected, which requires strict observation of the temperatures recommended by the manufacturer.

For bituminous materials modified in plant with a solid additive, special care should be taken in handling and proportioning that additive. In a continuous dryer drum mixer plant the additive is incorporated and proportioned by means of a special device (hopper + mechanical or pneumatic transfer and proportioning system). In a batch mixing plant the additive can be incorporated manually by introducing fixed amounts of it - from fusible bags having a predetermined weight - into the mixer, in proportion to the weights of the batches. There are also automatic devices for introducing and proportioning additives in batch mixing plants. Mixing sequences should be adapted to allow the additive to disperse completely and homogeneously in the bituminous material.

Laying temperatures should be observed as specified for each procedure. Some modified bituminous materials are very sticky, owing to the nature and/or high content of the binder. Compaction equipment and procedures should be adapted accordingly.

Fig. 6.12 Wearing courses in specific bituminous materials are generally rich in binder. This can be achieved by using additives (e.g., fibres) or modified bitumens.

6.3. Precracking

6.3.1 PRINCIPLE

The objective is to cause or induce transverse shrinkage cracking in a roadbase treated with cementitious binder, the induced cracks being spaced closely (generally 3 m) than those from on the natural shrinkage cracking process (generally 10 to 15 m). By increasing the number of cracks, it is hoped to reduce the width of cracks, thus leading to :

- less traction at the bottom of the wearing course when the cracks are opening in cold weather periods,
- smaller deflections and relative vertical movements of crack edges as heavy vehicles pass over, as a result of better interlock.

6.3.2 EXAMPLE OF PRECRACKING TECHNIQUES

Four procedures are mentioned here :

- injecting a fluid which prevents the setting of the cementitious binder : a transverse groove is made every 3 m. A fluid, generally a bitumen emulsion, is poured into it (figure 6.13),
- installing an element which reduces the section of the road base, thus serving as a crack inducer : this may be a plastic element or a plastic film (figure 6.14), for example,
- sawing a crack inducer, like in rigid pavements. The disadvantage of this technique is the need to operate after the material has hardened and yet before cracking has started : exactly how much time should be allowed is very variable and uncertain,
- cracking the pavement with a guillotine.

Fig. 6.13 Precracking machine injecting a fluid (in this case a bitumen emulsion) to inhibit cementitious setting.

Fig. 6.14 Precracking machine introducing a plastic film to reduce the section of the road base.

6.3.3 IMPLEMENTATION

Generally the techniques are implemented during the construction of the road base (see the first two procedures in Section 3-2). The sequence of operations is as follows :

- spreading the cementitious material,
- preliminary levelling with a grader, or spreading through a paver,
- precracking,
- preliminary compaction,
- fine grading with a grader or "autograde",
- final compaction.

On most construction sites, precracking can be fitted into the sequence easily. The only case which may present some difficulties is working lane by lane under traffic : two graders are then required.

6.4 References

1. F. VERHÉE : "Entretien des chaussées en béton de ciment. Bilan des idées actuelles et résultats des expérimentations françaises". Bulletin de Liaison des Laboratoires des Ponts et Chaussées, No. 97, September-October, 1975.

2. Proceedings of the 1st RILEM-Conference on Reflective Cracking in Pavements, Liège, 1989.

3. Proceedings of the 2nd RILEM-Conference on Reflective Cracking in Pavements, Liège, 1993.

4. Proceedings of the 3rd RILEM-Conference on Reflective Cracking in Pavements, Maastricht, 1996.

7

Summary and conclusions

A. Vanelstraete and L. Francken

Crack reflection through a road structure is one of the main issues of premature pavement deterioration. This is a widespread problem in many countries and highway maintenance authorities have to find economic means of repairing and upgrading their pavements.

There is evidence after several years of research and practice that materials and procedures having potential to improve the situation do really exist. But it must be emphasized that there is no standard solution suited for every situation. In most cases, improvements are the result of a combination of different functional layers into what is called "an overlay system". Successful solutions need a good expertise of each case and positive results can only be obtained if the right product is laid in the right way at the right place in the system and in the field.

7.1 Nature and origin of cracks (chapter 1)

Pavements are made up of layers and materials varying very widely both in nature and in properties. They are liable to fracture by many causes. The most important ones are : fatigue, thermal shrinkage, movements of the subgrade soil, constructional defects, ageing and environmental exposure. They give rise to cracks of highly different shapes and natures.

The reflection of an existing crack in an overlay is due to the fact that the edges of an existing crack are subject to movements which induce a concentration of stresses in the overlay. Traffic-induced stresses in a cracked pavement structure cause crack edges to move in mode 1 (opening), 2 (shear) or 3 (tearing), depending on the position of the vehicle with respect to the crack. These movements are rapid, frequent, and variable in amplitude. Thermal or drying shrinkage causes crack edges to move in mode 1 (opening). These movements are slow to very slow ; they are infrequent and great in amplitude.

The development of an existing crack, under the action of various loads and stresses, generally proceeds by three stages involving different mechanisms :

- in the initiation stage a crack is induced by a defect already present in the underlying layer,
- in the propagation stage the crack rises through the full thickness of the layer,
- the fracture or final stage is marked by the crack appearing at the surface of the layer.

Prevention of Reflective Cracking in Pavements. Edited by A. Vanelstraete and L. Francken. RILEM Report 18.
Published in 1997 by E & FN Spon, 2–6 Boundary Row, London SE1 8HN. ISBN 0 419 22950 7.

Any cracks appearing at the surface always have a detrimental effect on the pavement. There are ways to at least partly control the reflection of cracks from the underlayers, but the causes of these cracks and the shapes they may take are so varied that there cannot be any universal remedy for them. Before considering any solution to keep reflective cracking under control, one should, therefore, always make a correct and complete diagnosis of the problem to be solved.

7.2 Assessment and evaluation of the crack potential (chapter 2)

It is, indeed, vitally important to correctly diagnose the nature and causes of cracks in a structure to be treated, as it is this diagnosis which will direct the choice to proper solutions. Any procedure effective for certain types of cracking may be ineffective for others.

The first and inevitable stage in the evaluation of an existing pavement is the so-called problem identification phase. When this problem identification phase is completed and one has a first insight into the possible origins of the problems, a more quantitative and problem-specific evaluation can be performed. This second stage is the so-called quantification phase of the problem.

7.2.1 THE PROBLEM IDENTIFICATION PHASE

This phase involves the following steps :

7.2.1.1 Assessment of the type of road structure with its loading conditions

One of the basic requirements for the evaluation of the road structure is a knowledge of the type of road structure, with the thicknesses of the different layers, the type of layers and materials that were used, and the history of the road. It is also important to know the condition of drainage and, if possible, that of the base, subbase and subgrade. Some of this information can be provided after coring. Most important is a knowledge of the traffic and environmental conditions to which the road is exposed, with an estimation of the traffic to be expected in the future.

7.2.1.2 Visual Condition Surveys to identify the nature of the cracks

For overlay design purposes, crack mapping is an important tool, because this also helps to determine where the additional measurements can best be taken. It can be restricted to areas which are representative for the entire section and to areas where the damage is severe. Visual condition surveys also include a visual survey of the drainage system. Information is required on the level of the road with respect to that of its surroundings and on the presence or not of large slab stepping or other large unevennesses.

7.2.1.3 Coring

Cores taken on cracks can give essential information on the reflective potential of the crack and on the condition and thickness of the bound layers. When taking cores, it should be realised that not only the location where the crack is widest at the surface is of interest, but also cores taken at the tip of the crack is recommended to provide information on how the crack has developed. Several cores are sometimes necessary to determine how a crack has developed.

7.2.1.4 The assessment of the possible causes of the problems.

At the end of the problem identification phase one has a first insight into the possible origins of the problems, a more quantitative and problem-specific evaluation can now be performed.

7.2.2 THE PROBLEM QUANTIFICATION PHASE

The following measurements can be performed when the main causes are thermal and/or traffic related :

7.2.2.1 Falling weight deflection measurements (f.w.d) between the wheel paths

They are intended to back calculate the stiffness moduli of the different layers for further use in design models. They are taken between the wheel paths, because these areas have not been exposed to traffic.

7.2.2.2 Falling weight deflection measurements in the wheel paths

These are intended to derive the amount of damage due to traffic, which has resulted in loss of bending stiffness. These measurements should be taken in the wheel paths.

7.2.2.3 Deflection measurements at joints or cracks to analyse the load transfer

These measurements are especially important for the evaluation of concrete pavements to determine the load transfer across cracks. They can also be used to estimate whether or not voids or loss of support has developed due to e.g., erosion and pumping. From these measurements two types of approaches can be followed to determine the load transfer efficiency. Either they can be used on an empirical base with criteria for good, medium and low load transfer; or they can be used in finite element analysis.

7.2.2.4 Crack activity measurements - Slab rocking displacements

They are interesting for cement concrete slabs, to determine slab rocking. Although measurements of slab rocking are highly dependent on temperature and on the humidity conditions of the subbase and subgrade, they can give indications whether or not the concrete slabs have to be cracked and seated before overlaying. The results of these measurements can also be used in finite element analysis to obtain information about the load transfer across cracks.

7.2.2.5 Crack width measurements in relation to temperature
In some cases of very severe cracking, crack width measurements together with measurements of temperature and moisture can be performed. These are, however, more used for scientific purposes.

7.3 Crack prevention and use of overlay systems (chapter 3)

Two categories of methods are currently used to reduce the appearance of cracks at the road surface. The first category concerns methods to prevent or reduce the formation of cracks during the initial construction phase or to treat existing cracks which induce less severe cracking or make cracks less active. The second category of methods concerns the use of an overlay system. This term is used to describe the combined system of a bituminous overlay, interlayer system and levelling course, placed on an underlying road structure.

7.3.1 PREVENTION METHODS AND TREATMENT TECHNIQUES FOR CRACKS BEFORE OVERLAYING

A correct choice of the base materials, a proper design of the road structure, and a good quality of placement are basic rules to respect :

- For cement treated bases, it is recommended to use aggregates with low coefficients of thermal expansion.
- Also for cement treated bases, different precracking methods exist and have proven their effectivity. They are inexpensive and can be implemented without many constraints.
- For bituminous mixes, the use of certain polymers or additives can improve their cracking resistance.
- If the origin of the cracks is known, techniques exist in some cases to eliminate the origin of existing cracks before overlaying, e.g., by drainage and by sealing the surface in case of excessive moisture content, structural strengthening in case of fatigue, planing and placement of a new course with a good bond to the underlying layer in case of slippage.
- Crack and seat techniques on cement concrete slabs in the case of large vertical movements measured at the joint or crack edges to limit the activity of cracks. Evidence of the long term benefit of this technique has been provided. This technique, however, entails a certain loss of bearing capacity and must, therefore, be decided on the basis of a careful analysis completed by a proper structural design.
- Sawing of cement concrete slabs.
- Sawing and sealing of the overlay above the joints in the cement treated bases.
- Injection techniques preventing water to penetrate to the underlying structure.
- Bridging of cracks with an elastomer-bitumen film to restore watertightness.

7.3.2 THE USE OF OVERLAY SYSTEMS

Special care is necessary in determining the design and the characteristics of the asphaltic overlay itself. In addition, an appropriate interlayer system may be used between the old structure and the new overlay to slow down the initiation and propagation of cracks in the overlay.

Although the range of commercially available interlayer products is very wide, the large variety of products can be classified in a limited number of categories. Often used are sand asphalts, SAMIs, nonwovens, grids and steel reinforcement nettings. Different fixing layers or methods, e.g., tack coat, binder layer, slurry seal or nailing, are used to provide a good adhesion of the interlayer product to the underlayer. The choice of a given fixing method/layer depends on the type of interlayer product.

The role of an interlayer system in the road structure depends on the type of interlayer system and can be :

- To take up the large localised stresses in the vicinity of cracks and, hence, reduce the stresses in the bituminous overlay above the crack tip. The product in that case acts as reinforcement product. This is the case for most grids and steel reinforcing nettings (see also 7.4).
- To provide a flexible layer able to deform horizontally without breaking in order to allow the large movements taking place in the vicinity of cracks. This is the case for impregnated nonwovens, for SAMIs and for sand asphalts. This function is also often described as "controlled debonding". It is obvious that total debonding has to be avoided in all cases, otherwise fatigue cracking or delamination may appear already very shortly after rehabilitation.
- To provide a waterproofing function and keep the road structure waterproof even after reappearance of the crack at the road surface. This is the case for nonwovens and SAMIs.

Interlayer products are unable in any case to prevent the movements of cracks or joints already present in the base layer, whatever the bonding system.

The most important component of an overlay system remains the bituminous layer itself. By its thickness and performance it will definitely influence the service lifetime of the road, even with the use of an interlayer system. Indeed, a thicker overlay greatly reduces the stresses induced by traffic at the existing cracks (initiation stage). A thicker overlay also lengthens the path to be followed by an incipient crack before it reaches the surface (propagation stage). As for thermal cracking, a greater thickness offers a protection which mitigates differences in temperature, thus reducing the amplitude of movement of crack edges.

Although the overlay thickness has a strong influence on the time when reflection cracks reappear, it was observed for thick overlays that cracks often initiate at the top of the bituminous layer and propagate downwards to meet the crack in the underlying layer. In these cases, the onset of reflective cracking is largely determined by the properties of the wearing course.

The resistance of a bituminous material to cracking depends on the mix design and on the characteristics of the materials. The use of polymer modified binders and fibres has been found effective to reduce reflective cracking.

7.4 Characterization of overlay systems (chapter 4)

In order to choose a suitable overlay system for a given situation, the characteristics which are relevant for the role they have to play in the road structure have to be determined.

Characterization of interlayer systems implies that tests are performed on the interlayer product as well as on the interlayer system as part of the overlay system.

7.4.1 CHARACTERIZATION OF INTERLAYER PRODUCTS

For nonwovens, grids and steel reinforcing nettings, the relevant properties are :

- base material : polyester, polypropylene, fibre glass or steel,
- thickness (in mm),
- rigidity of the junctions,
- mesh width : this is only relevant for grids and steel reinforcing nettings,
- ultimate strength (in kN/m),
- strain at ultimate strength (in %),
- product stiffness (in kN/m),
- stiffness modulus of the interlayer material (in MPa),
- temperature susceptibility; the product may not deform or melt during placement of the asphaltic overlay,
- quantity of bitumen that can be absorbed by the interlayer product (is only relevant for nonwovens).

Simple tensile tests are generally performed on the interlayer products to deduce the ultimate strength, strain at ultimate strength, product stiffness and stiffness modulus. These values are often anisotropic. Different testing methods exist in the different countries and the testing methods also depend on the type of products. Hence, even for two types of grids, the specified values are not always comparable.

From their mechanical properties, the role that interlayer systems play in the road function can be deduced. Nonwovens are characterized by a low stiffness modulus, compared to that of the bituminous mix, and are therefore not relevant as reinforcement products. However, they have a high strain at ultimate strength, which makes them suitable to withstand large horizontal deformations as there exist just above the crack tip, e.g., in the case of temperature variations. Grids and steel reinforcing nettings are characterized by a high stiffness modulus. They act as reinforcement products.

In order to determine whether or not a given product acts as reinforcement in a given situation, the stiffness modulus of the interlayer product has to be compared with that of the overlay. Taking into account that the overlay stiffness modulus is highly temperature dependent, and changes also with frequency and with the lifetime of the overlay, a given interlayer system can be reinforcing in one situation and not in another situation (e.g. summer versus winter; propagation phase versus initiation phase). This is the case for some grids.

7.4.2 CHARACTERIZATION OF OVERLAY SYSTEMS

Laboratory tests to study the effect of interlayer systems as part of an overlay system concern mainly two types of tests :

- Tests to determine the adherence of the interlayer system with the underlayer and overlay. The adherence depends on the fixing method/layer and the used quantity, but also on the type of interlayer product. Simple tensile tests, pullout tests and shear tests are generally used for this purpose.
- Laboratory tests to investigate the performance of overlay systems. Testing facilities exist to study their effect in case of thermal loading (shrinkage of the base layer, thermal shrinkage of the overlay), traffic loading (with presence or not of large vertical movements at the crack edges) and combined traffic and thermal loading.

From these laboratory tests, it was found that efficient crack retarding systems do exist, however, not in all cases. A lot of interlayer systems are effective in case of horizontal (mode I) crack opening. For nonwovens and SAMIs, their performance is mainly determined by the amount and type of binder they contain.

With the available test equipments, a qualitative ranking of interlayer systems, under well-defined test conditions, is now possible. Although there seems to be a general agreement between laboratory results and field experiments, a lot of laboratory results have however never or seldom been adjusted to full scale results. And there are even less cases where this adjustment is made on the basis of a statistical analysis from several test sites.

7.5 Modelling of overlay systems (chapter 5)

Models and software tools specifically developed to study the phenomenon of reflective cracking exist and are still being improved. Modelling of overlay systems allow a better insight into the behaviour of overlay systems and may suggest improvements in design. Common to all models is that they are constituted by a computing technique to determine the stress-strain distribution in the structure and a set of physical deterioration laws describing the behaviour of the structure and its evolution under service conditions. For the initiation phase of a crack in the overlay, the lifetime of the overlay is determined via computation of the tensile strain at the bottom of the overlay and subsequently making use of a fourth power fatigue law. For the propagation phase of the crack through the overlay, analyses are performed by using Paris's law.

The use of the finite element method for structural design calculations has become very popular during the last years. Although a lot of interesting work has been done, they depend - and will always depend - on the quality and relevance of the assumptions and the quality and accuracy of the input data. Indeed, numerous testing methods exist for determining the materials characteristics used in the physical laws which are applied. However, there is a lack of information on the influence of mix composition and on their temperature dependence and loading time susceptibility.

Paris's law is the only tool we have to predict crack propagation, although we know that the conditions of homogeneity, isotropy and linearity are not fulfilled. Moreover, test results obtained with this law generally present a large scatter.

Large efforts are needed in the field of modelling to come to real practical solutions for the design of overlay systems. Only tentative results are given so far, describing particular situations under specific loading conditions. It is not clear whether the results obtained by different authors are comparable and can be extrapolated to other cases. The situations being studied so far either concern the crack initiation or the crack propagation phase, but do not describe the whole cracking phenomenon.

7.6 Implementation of overlay systems on the construction site (chapter 6)

Careful installation of interlayer systems is predominant for a good performance of these systems, with specific rules to respect for each type of interlayer product. Even the best solution for a given situation can turn into a disaster if some of the basic rules of good practice are not carefully followed on the site. These rules must be clearly defined, validated and implemented in standard tender specifications.

The laying procedure of interlayer systems generally comprises the following consecutive stages : preparatory works, the application of a fixing layer, the application of the product itself, the placement of a protective layer (for grids) and the application of the bituminous overlay. Rules have to be respected at each stage, some examples are:

- Concerning preparatory works :
  A levelling course might be necessary in order to place the product on a flat surface. Cracks wider than 5 mm should be sealed to waterproof them and hold their edges together. On cement concrete pavements subjected to severe vertical movements at joints, it may be desirable to reduce those movements. When slab rocking is larger than 0.50 mm crack and seat techniques have to be used. However, one has to take into account that cracking and seating the concrete slabs gives rise to a certain loss in bearing capacity. This has to be taken into account for the structural design of the road. When slab rocking is of the order of 0.25 to 0.50 mm, the relative vertical movements can be mitigated by improving load transfer across the joints (by installing dowels or connectors at transverse joints) or by reducing

the cavities under the slabs by filling with a cementitious slurry. It is absolutely necessary to avoid cavities under the interlayer product. These may remain present after overlaying or/and result in locally inadequately compacted spots.

- For the fixing layer:
  In order to achieve an overall and homogeneous fixing of the interlayer product to the underlayer, it is recommended to spread out the fixing layer before unrolling the interlayer product, except for steel reinforcing nettings which are embedded in a slurry being applied after placement of the interlayer product.
- For the application of the interlayer product :
  To prevent problems during overlaying, any folds, creases or wrinkles, which are unavoidable when the product is unrolled on bendy roads, must therefore be cut, after which the edges should be pressed down over each other and fixed.
  No traffic must be allowed on grids before the bituminous overlay is in place, except for site traffic laying the overlay. These vehicles should avoid braking, accelerating, or turning abruptly on the grid. In some countries specifications require a protective layer to reduce the risks of displacement or wrinkling under site traffic.
- For the bituminous overlay it is recommended that the overlay thickness is minimum 3 cm on nonwovens and 4 cm on grids and steel reinforcing nettings.

7.7 Remaining issues

The success of innovative solutions depends on the correct choice of all the components of the overlay system, on their combination and on their implementation in function of the loading conditions to which they will be exposed for a future design life. Typical questions to be answered by road designers are : which surfacing thickness is needed in a given situation ? What is the relative lifetime of a given solution in comparison with a classical solution ? At this moment, these questions can only be answered with big uncertainties. With the present knowledge from laboratory tests, modelling and in field experience, it is very difficult to predict the improvement in service lifetime of a given interlayer system on the field, for any possible situation. With the available test equipments, a qualitative ranking of interlayer systems, under well-defined test conditions, is now possible. Although there seems to be a general agreement between laboratory results and field experiments, a lot of laboratory results have, however, never or seldom been adjusted to full scale results. Also, they are usually performed under ideal laying conditions, whereas it has been observed worldwide that a lot of projects already failed in a very early stage, because of bad placement. Large efforts are still needed in the field of modelling to come to real practical solutions for the design of overlay systems. Only tentative results are given so far, describing particular situations under specific loading conditions.

Some other important problems deserving more attention were hardly dealt with. How to reclaim and recycle interlayer systems ? What is the long-term performance of interlayer products ? To what extent are they affected by chemical reactions or corrosion ?

Limited information is also available on the long-term performance of different overlay systems. These remain all interesting topics to be studied in the future.

Index

Adherence tests 61
Assessment 16

Bituminous overlay 43
Blunt crack band theory 84

Characterization 61
Combination products 43
Construction site 104
Controlled debonding 43
Crack activity measurements 16
Crack path 1
Crack potential 16
Crack prevention 43
Crack propagation 16
Crack propagation 84
Crack reflection 1
Cracking 1
Cracking forms 1
Cracking types 1

Deflection measurements 16
Detrimental effects 1

Environmental induced cracking 16
Equilibrium equations 84

Finite difference method 84
Finite element method 84
Fixing layer / method 43
Fracture mechanics 84

Grid 43
Grids 104

Implementation 104
Interlayer product 43
Interlayer product 61
Interlayer system 43
Interlayer system 61
Interlayer systems 104

Load transfer 16
Loading conditions 84
Loads 1

Mechanistic empirical overlay design method 84
Membrane with threads 104
Modelling 84
Movements at cracks 1
Multilayer linear elastic 84

Nonwoven 43
Nonwovens 104

Origin of cracks 1
Overlay system 43
Overlay system 61
Overlay systems 104

Pavement evaluation 16
Performance laws 84
Precracking 104
Precracking 43
Preparatory works 104
Pull-out 61
Reinforcement 43
Road structures 1

SAMI 104
SAMI 43
Sand asphalt 104
Sealing 104

Shear tests 61
Simulation of thermal loads 61
Simulation of traffic 61
Slab rocking 104
Steel reinforcing nettings 104
Steel reinforcing nettings 43
Stiffness 61
Strain 61
Strength 61
Stresses 1
Structural design 84

Tensile tests 61
Thermal stress 16
Traffic induced cracking 16

Watertightness 43

Printed and bound by CPI Group (UK) Ltd, Croydon, CR0 4YY
01/11/2024
01782621-0018